American Institute of Chemical Engineers

Topical Conference on Distillation 2007

Held at the 2007 AIChE Spring National Meeting

April 22-27, 2007
Houston, Texas, USA

Printed from e-media with permission by:

Curran Associates, Inc.
57 Morehouse Lane
Red Hook, NY 12571
www.proceedings.com

ISBN: 978-1-60423-355-1

Some format issues inherent in the e-media version may also appear in this print version.

ISBN: 978-1-60423-355-1

American Institute of Chemical Engineers
Proceedings
Three Park Avenue
New York, NY 10016-5991
Phone: 212-591-8100

www.aiche.org

American Institute of Chemical Engineers

Topical Conference on Distillation
2007

TABLE OF CONTENTS

Session 6 – Energy Savings and Process Innovations: Continuing the Heritage of Kunesh, Sakata and Zuiderweg

Session 28 – Advances in Refinery Distillation: Continuing the Heritage of Kunesh, Sakata and Zuiderweg

Session 54 – New and Exciting Dynamics and Control Applications: Continuing the Heritage of Kunesh, Sakata and Zuiderweg

Session 78 – Standard and High-Capacity Trays: Continuing the Heritage of Kunesh, Sakata and Zuiderweg

Session 98 – New and Exciting Packing Applications: Continuing the Heritage of Kunesh, Sakata and Zuiderweg

Session 107 – Distillation Troubleshooting: Continuing the Heritage of Kunesh, Sakata and Zuiderweg

Session 120 – New Processes and Enhanced Distillation: Continuing the Heritage of Kunesh, Sakata and Zuiderweg

Energy-Saving Distillation through Internal Heat Exchange (HIDiC): Overview of a Japanese National Project

Shuzo Ohe, Science University of Tokyo
1-3 Kagurazaka, Shinjuku-ku, Tokyo, Japan 162-8601

INTRODUCTION

Nowadays, the world is faced with significant energy problems. High oil prices are forcing a need for energy reduction. Global warming must be headed off by reducing CO_2. For these and many other reasons, energy saving is a practice that must be pursued constantly by everyone.

In distillation too, energy conservation has long been a key problem. Separating mixtures by distillation requires a great deal of heat energy. In conventional systems, many approaches have been tried, but with limited results. Using recent energy-saving technology, the industry has been able to reduce the amount of heat energy consumed in a large-scale chemical plant, compared to earlier operations. However, the capability of that technology has reached its limits.

In March 2006, the Japanese government concluded the HIDiC project, a four-year project on energy saving in distillation. HIDiC stands for Heat Integrated Distillation Column. This innovative system offers some strong and unusual advantages. As chairman of the project review committee, I would like to provide here a brief summary of the system's principles, as well as some of the project's goals and results.

The HIDiC system employs a breakthrough concept in the distillation process that can deliver a new, higher capability for saving energy. As a result, HIDiC can offer more cost savings than any previous system. (For example, the overall energy savings target for the project was 30%, but the actually demonstrated savings were 60% for the petrochemical plant and 40% for the air separation plant.)

The striking difference between HIDiC and predecessor systems is in the way that HIDiC handles heat. In the conventional distillation system, the waste heat at the top of the enriching section is not recovered. This makes the heat utilization of the conventional system comparatively low. The HIDiC system, in contrast, makes significantly more utilization of waste heat.

HIDiC makes use of the heat from the enriching section to heat the stripping section. This is made possible through the HIDiC system's distinctive longitudinal partitioning of the distillation column into its enriching and stripping sections. In this approach pressure is applied to the vapor, compressing it and causing the temperature to rise, and the resultant heat is thus transferred to the stripping section.

As a result, unlike conventional distillation systems, HIDiC requires no external reflux. That is because, by transferring heat from the enriching section to the stripping section, the HIDiC system can make its own internal reflux.

1. PRINCIPLES

HIDiC adapts a principle which Mah et al. originally presented about thirty years ago as SRV (Secondary Reflux and Vaporization) distillation. (The SRV column is also cited in Perry's Handbook.) HIDiC manifests this concept industrially to achieve heat transfer from the enriching section to the stripping section. It does this by applying SRV and eliminating reflux at the top of the enriching section. The stages of the process are as follows.

1. Vapor from the stripping section does not go directly to the enriching section. Instead, it goes there after compression raises the vapor temperature.

2. Then heat effectively transfers from the vapor in each enriching section tray to the liquid in the matching stripping section tray.

3. The pressure in the enriching section is higher than in the stripping section. After the pressure of the liquid from the enriching section is reduced, the liquid is then fed to the stripping section.

4. Then the vapor rate in the enriching section decreases progressively as it approaches the top. At the same time, the liquid rate in the enriching section increases progressively as it approaches the bottom of the section.

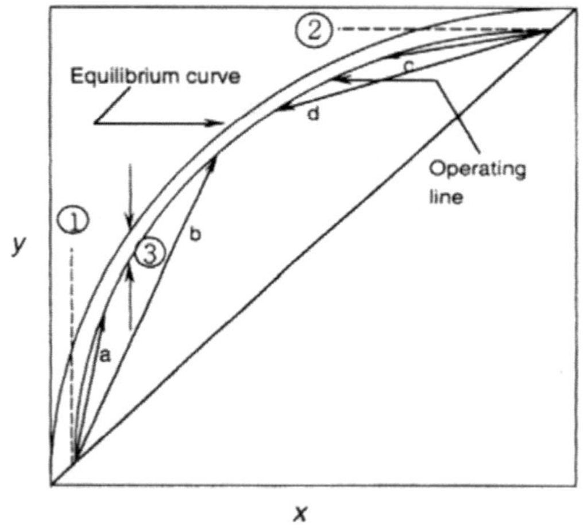

Figure 1. Conditions Required for Internal Heat Exchange[1]

[1]Takamatsu, B., M. Nakaiwa, T. Nakanishi, *Kagaku Kogaku Ronbunshu,* **22**, 985 (1996)

The SRV principle originated by Mah et al. was the springboard for the HIDiC principles developed by Takamatsu, Nakaiwa, and Nakanishi. Their HIDiC concept is illustrated here in an x-y diagram. They asserted that the implementation of a HIDiC system requires all three of the following conditions:

1. The operating line in the *stripping* section is *vertical* at the *bottom* of the column.
2. The operating line in the *enriching* section is *horizonta*l at the *top* of the column.
3. The distance between the operating line and the equilibrium curve is an appropriate value throughout the column.

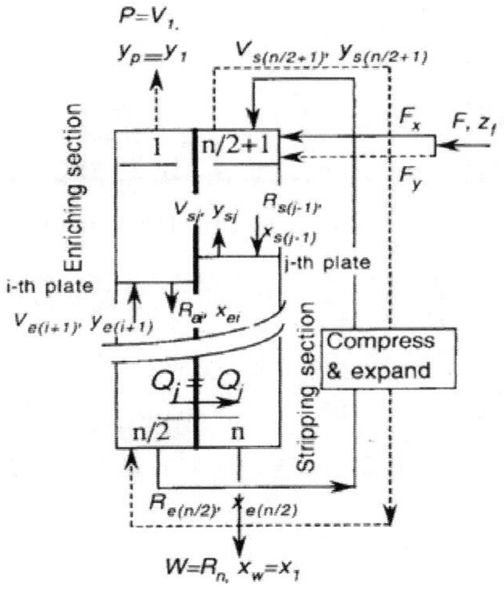

Figure 2. Principles on Which HIDiC is Based[1]

[1]Takamatsu, B., M. Nakaiwa, T. Nakanishi, *Kagaku Kogaku Ronbunshu,* **22**, 985 (1996)

Figure 2 shows the relationships of calculations for flow rate and composition involved in the HIDiC system. The feed, F, enters at the top of the stripping section, as marked by the left-facing arrow, and goes to the stripping section. The vapor, V, from the stripping section is compressed and fed to the enriching section. The top product, P, exits the top of the enriching section.

The mass balances are given by equations (1) through (3). One of them can also define the q-value. Based on the amount of Q transferred from the enriching section, equation (4) can calculate the amount of V in the stripping section.

Mass Balances

$$F = P + W, \quad Fz_f = Py_p + Wx_w \tag{1}$$

$$W / F = q \tag{2}$$

$$z_f = (1-q)y_p + qx_w \tag{3}$$

Action of Internal heat exchange

$$R_{e(n/2)} = V_{s(n/2+1)} - P = V_{s(n/2+1)} - F_y$$

$$= \sum_{i=1}^{n/2} Q_i / \lambda = \sum_{j=n/2+1}^{n} Q_j / \lambda = Q / \lambda \tag{4}$$

The operating line of the enriching section is given by equation (7). In this equation, R_{ei} is smaller, so the slope $R_{ei}/R_{e(i+1)}$ becomes smaller. This means that a condenser is not needed. However, internal reflux is needed. Therefore internal heat exchange is required for reflux of each plate.

i-th plate in enriching section

$$V_{e(i+1)} = R_{ei} + P \tag{5}$$

$$V_{e(i+1)}y_{e(i+1)} = R_{ei}x_{ei} + Py_p \tag{6}$$

$$y_{e(i+1)} = \frac{R_{ei}}{V_{e(i+1)}}x_{ei} + \frac{y_p}{V_{e(i+1)}} \tag{7}$$

The operating line of the stripping section is given by equation (10). In equation (10), $V_{s(j-1)}$ is smaller, so the slope $R_{s(j-1)}/V_{s(j-1)}$ becomes larger. This means that a reboiler is not needed. However, internal vaporization is needed. Therefore internal heat exchange is required for vaporization of each plate.

j-th plate in stripping section

$$R_{s(i-1)} = V_{sj} + W \tag{8}$$

$$V_{s(i-1)}y_{sj} = R_{s(j-1)}x_{sj} - Wx_w \tag{9}$$

$$y_{sj} = \frac{R_{s(j-1)}}{V_{s(j-1)}}x_{s(j-1)} - \frac{x_w}{V_{s(j-1)}} \tag{10}$$

The operating line equations (11) and (12) are shown here. The internal reflux ratio and internal reboiling ratio are also given.

Enriching section

$$y_{e(i+1)} = \frac{re_i}{1+re_i}x_{ei} + \frac{y_p}{1+re_i} \quad \text{Internal reflux ratio}: re_i = \frac{R_{ei}}{P} \tag{11}$$

Stripping section

$$y_{sj} = \frac{1+rs_j}{rs_j}x_{s(j-1)} - \frac{x_w}{rs_j} \quad \text{Internal vaporization ratio}: r_{sj} = \frac{V_{sj}}{W} \tag{12}$$

The internal reflux ratio is different at each plate. The ratio is expressed as the product of *i* and dr_e, in equation (13) or the product of (n-j+1) and dr_s in equation (14). *The values for* dr_e and dr_s are determined from the amount of internal heat transfer.

Enriching section

$$y_{e(i+1)} = \frac{idr_e}{1+idr_e}x_{ei} + \frac{y_p}{1+idr_e} \qquad re_i = i \times dr_e \tag{13}$$

Stripping section

$$y_{sj} = \frac{1+(n-j+1)dr_s}{(n-j+1)dr_s}x_{s(j-1)} - \frac{x_w}{(n-j+1)dr_s} \qquad rs_j = (n-j+1)dr_s \tag{14}$$

$$dr_e = Q_k/\lambda/P, \qquad dr_s = Q_k/\lambda/W \tag{15}$$

2. PILOT PLANT PROJECT

HIDiC is evaluated highly as a chemical process in the international effort to combat global warming. In Japanese industry as a whole, 15% of all energy is expended in the chemical industry. In the chemical industry, separation by distillation expends 40% of total energy. The basic research on HIDiC has already yielded promising results. However, significant issues remain before profitability can be assured. Much additional funding is required for its development, but support from private enterprise cannot be expected. Hence, the project must be conducted by NEDO, an agency of the Japanese government.

For the basic study, a project was performed (1993-2000), with encouraging results. The energy saved was calculated as more than 30% over that of the same kind of plant operating with a conventional system. Moreover, the specifications were maintained without reflux – a condition essential for ordinary distillation – for five full hours.

The project focused on the tray, structured packing and plate fin column systems in distillation processes of the petrochemical and air separation industries. The term was four years (from April 2002 to March 2006) and the budget was 1,400,000,000 yen (about US $12,730,000 at that time).

The project was broken down into five portions. For each portion, highly competent leaders were selected from companies with reputations for expertise in research and development in the subject areas.

The budget was distributed according to the objective of each portion. One third of the budget was allocated to the construction and operation of the pilot plant. The distributions for the three types of HIDiC distillation systems (investigating shell and tube, tray, and plate fin) were also sufficient, taking into consideration the cost of making the test equipment.

The four-year schedule was divided into two terms of two years each. The first term was used for basic study and the second term was used for pilot plant construction and testing to evaluate the project. Special care was given to designing a cost-effective and expedient project schedule.

Between the two terms, evaluation of the competing systems was performed by an independent committee comprised of members from industries and academia. This evaluation showed that the principle of competition was promoting efficient use of budget and time.

To optimize the HIDiC design, especially for large-scale distillation columns, the following items will require further R&D.

1. Achievement of both homogeneous distribution of vapor and liquid flow and effective heat transfer through the wall.
2. Development of technology for multi-component systems (ternary and greater).
3. Design of a structure that precludes the complex problems that can arise when the up and down flow rates differ greatly.
4. Development of precise pressure control technology to enable practical application of the azeotropic mixture.

The HIDiC pilot plant (Fig. 3) is located in the factory of Maruzen Petrochemical Company, Limited, near Tokyo, Japan.

Figure 3. HIDiC Pilot Plant

Figure 4 shows the compressor for pressurizing the vapor from the stripping section before it enters the enriching section.

Figure 4. Compressor for the Pilot Plant

This sign (Fig. 5) lists the governmental agencies and participating companies involved in the design, construction and operation of the HIDiC pilot plant.

Figure 5. HIDiC Project Participants

Figure 6 shows some of the people connected with the project. Kneeling in the front row on the far left is the factory director, Mr. Iuchi, then the five members of the Review Committee. The author is second from the right, and project team members are on either side. Standing fourth from the right you can see the project leader, Dr. Nakaiwa.

Figure 6. Project and Review Committee Members

Outline of the Project

The basic study, HIDiC 1, extended from 1993 to 2000. Two major results were achieved: energy savings of more than 30%, and maintaining of process specifications for five hours without reflux. The basic project focused on the tray, structured packing, and plate fin column systems in distillation

HIDiC 2, the advanced project, ran from 2002 to 2006. The budget was one trillion, 400 billion yen (about 14 million U.S. dollars). The project was broken down into five portions. One third of the budget was allocated to the construction and operation of the pilot plant.

Project Budget and Time Schedule

Table 1 shows the budget and schedule of the project. The project started in 2002 and concluded at the end of the 2005 Japanese fiscal year i.e., the end of March 2006. The budget is shown for each year, in millions of yen and millions of dollars. The total budget used was $1,360,000, at a rate of 100 yen to the dollar.

Table 1. Project Time Schedule and Budge

Schedule	Fiscal year	2002	2003	2004	2005	
	Petro-Chemical	—	—	—	—	
	Air Separation	—	—	—	—	
	Multi-component	—	—	—	—	
	Report				—	
Budget	Fiscal year	2002	2003	2004	2005	Total
	Million yen	220.0	343.2	572.4	232.2	1367.7
	Million dollar	2.200	3.432	5.724	2.322	13.677

Project Team

Figure 7 is the project's organization chart. The project was organized by a Japanese government entity, NEDO (the New-Energy and Industry Technology Development Organization). The project leader was Dr. Nakaiwa, who belongs to AIST, a Japanese government technology institute. There were two group leaders. One was Mr. Iuchi, director of Maruzen Petrochemical Company. The other was Mr. Kawakami, manager of Taiyo Nissan. Five companies took part in this project.

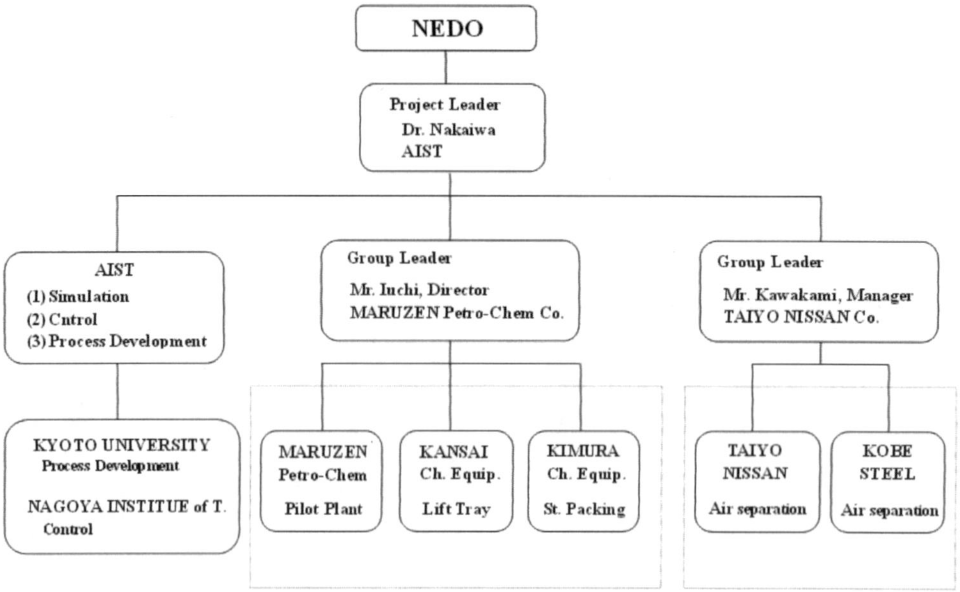

Figure 7. Project Organization Chart

Review Committee

The six-member independent Review Committee listed in Table 2 was composed of representatives from three industrial companies and professors from three universities. At the conclusion of the project, the Committee's results were presented at a meeting of 47 individuals from Japanese government agencies (METI, the Ministry of Economy, Trade and Industry; and NEDO), and the companies that received government funding and took part in the project.

Table 2. Review Committee Members

Position	Name	Title, Affiliation
Chairman	Shuzo Ohe	Professor, Science University of Tokyo
Vice Chairman	Yuji Naka	Professor, Tokyo Institute of Technology
Member	Shigeru Koyama	Professor, Kyushu University
Member	Yoshitaka Saito	Manager, Alcohol Association
Member	Kazuo Matsuda	Manager, Chiyoda Engineering Company
Member	Shizuo Midori	Consultant, Yokaichi Petrochemical Company

3. PILOT PLANT CONSTRUCTION AND PERFORMANCE

HIDiC Pilot Plant Specifications

The structure of the HIDiC is that of a shell and tube heat exchanger. The shell side is used as the enriching section, while the tube side is used as the stripping section. Both sides are filled with structured packing. Dimensions are shown in this table. The column diameter is 1.4 m and the inner tube diameters are 267 mm, 216 mm and 165 mm. The size difference is based mainly on the difference in the vapor rate in the tube as heat is transferred to the shell side.

Table 3. Comparison of the Pilot Plant with a Commercial Plant

		HIDiC	Conventional
1	Column		
	Height	27 m	31 m
	Diameter	1.4 m	1.0 m
	Internal	Structured Packing 13.5 m H.	Bubble Cap 55 trays
2	Compressor	45 KW	-
3	Reboiler	7.4 m^2	23 m^2
4	Preheater	4.8 m^2	-
5	Condenser	23.5 m^2	87 m^2

HIDiC Structural Details

Figure 8 shows the internal structural details of the HIDiC unit. Each tube is an independent column. The enriching section is in the tube side and the stripping section is in the shell side.

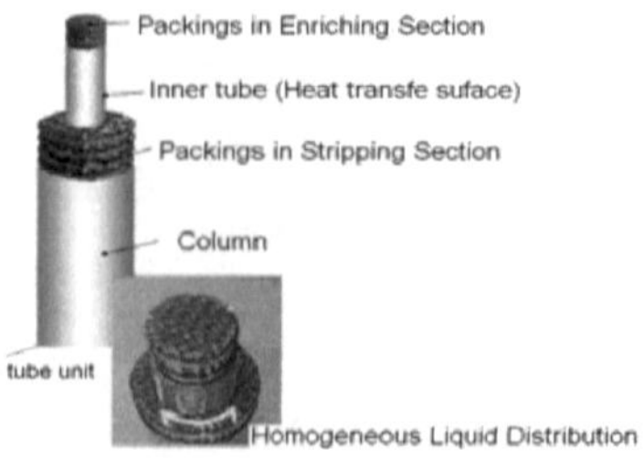

Figure 8. HIDiC Internal Structural Details

In Fig. 9 you can see the structure of the HIDiC as it was actually constructed. Each tube was inspected in the factory. Then it was mounted in the column. The column was then shipped to the pilot plant site.

Figure 9. Inspection, mounting and shipping of HIDiC unit

The distillation efficiencies were satisfactory for both sides or sections. Figure 10 shows the improvement in efficiency after the project. The NSTM number of theoretical plates per unit height of packing bed, originally ranging from one to two point five, increased to a constant of three.

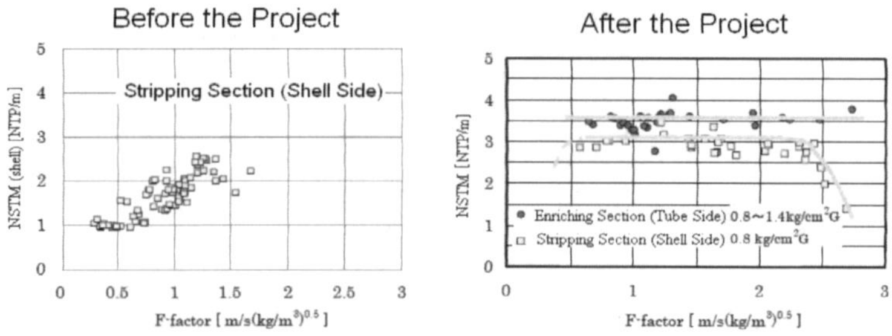

Initial efficiency After modification

Figure 10. Improved Efficiency after Modification

Testing confirmed the capability of continuous operation for 1000 hours. Table 4 shows the results for three tests. While the average showed slightly more than 60% improvement in energy savings, the individual test results were each considerably higher.

Table 4. Energy Savings Results

Energy Saving Result

Date	Feed Rate (kg/hr)	External Reflux Ratio	Energy Saving (%)
Nov. 11, 2005	1900	0	63.9
Feb. 2, 2006	1700	0	67.0
Mar. 10, 2006	2100	0	63.5
Average (year)	1714	1.5	61.2

The chart in Fig. 11 shows the control performance. After the feed rate was changed, steady states were achieved within 3 hours. A steady state was maintained for each flow rate — reflux, bottom, and distillate. This demonstrates HIDiC's exceptional stability in operating performance.

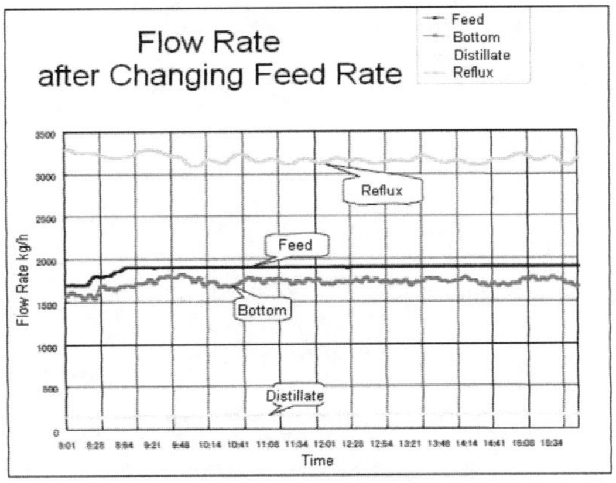

Figure 11. Flow Rate After Hanging Feed Rate

Lift Tray for HIDiC Application

Testing was also done to evaluate the suitability of the tray type for application in the HIDiC. The lift tray was adopted because of its flexible operation. Interior and exterior views of the lift tray are shown in Fig. 12.

Each tray is composed of a pair of perforated plates. The upper perforated plate can be adjusted up or down, depending upon the duty required. The tray is a type of counter flow, so no downcomer is required. This is advantageous for HIDiC, as it allows greater direct contact of the vapor with the shell side liquid.

Figure 12. Lift Tray

Details of the Lift Tray Column

The details of the tested lift tray are shown in Fig. 13. The enriching section is on the outside and the stripping section is on the inside.

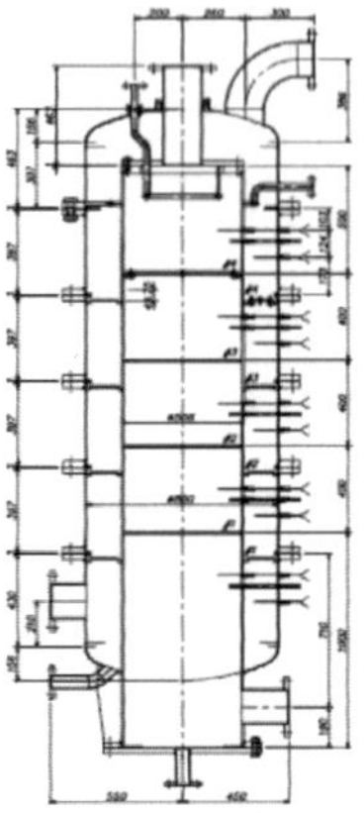

Inside Column Diameter 508 mm OD

Outside Column Diameter 800 mm ID

Column Height 2640 mm

Surface Area of Inside Column 3.97 m2

Numbers of Plates 4

Figure 13. Lift Tray Details

4. DISCUSSION

HIDiC's innovative design principles eliminate the need for external reflux. Here we will discuss the reasons for this.

NTP of Ideal HIDiC

First, let us examine the calculation of the number of theoretical plates (NTP). Figure 14 is reproduced with the permission of the original authors[1]. Note that at the top, the operating line is almost horizontal, whereas at the bottom it is vertical.

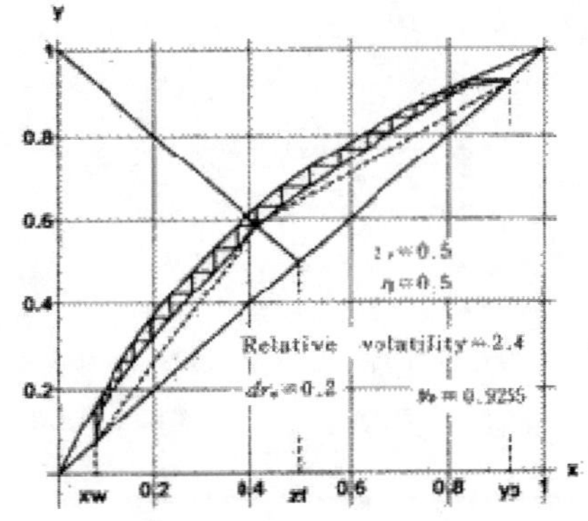

Fig. 6 A numerical example of No. of theoritical plates calculation in the ideal HIDiC for benzene-toluene system.

Figure 14. NTP of Ideal HIDiC

[1]Takamatsu, B., M. Nakaiwa, T. Nakanishi, *Kagaku Kogaku Ronbunshu,* **22**, 985 (1996)

We performed calculations based on the method described in the paper by Takamatsu et al. The result is shown in Fig. 15.

The result that we obtained coincides 100% with that of the original authors as shown in Fig. 14.

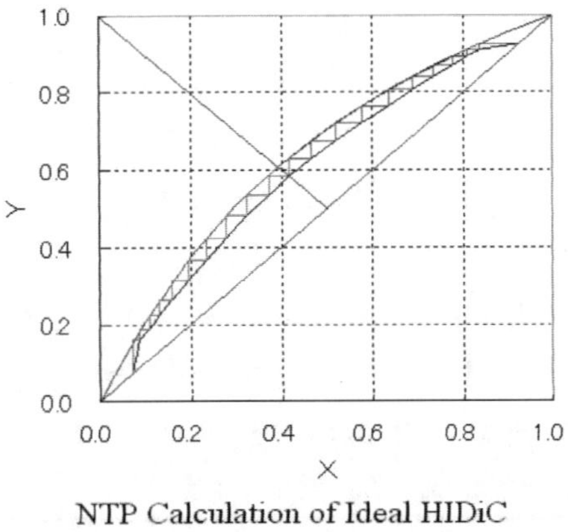

Figure 15. NTP of Ideal HIDiC

Using bubbling point calculations, we calculated the NTP for the differing pressures of the stripping and enriching sections. These results are shown in Fig. 16.

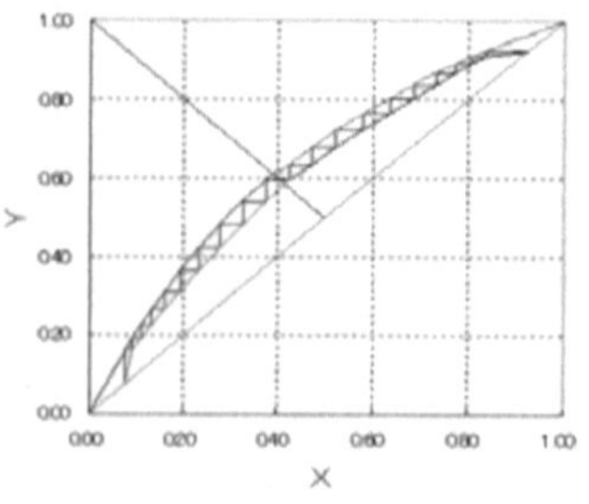

NTP Calculation of Ideal HIDiC (Bubble and Dew point Calculation)

Figure 16. NTP for the Differing Pressures

Figure 17 is the temperature profile for the differing pressures in the stripping section (760 mm Hg) and the enriching section (1520 mm Hg). The difference of temperature between the sections is about 20°C.

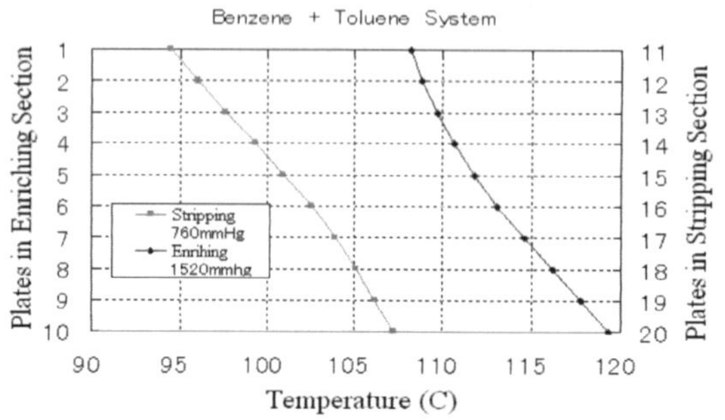

Figure 17. Temperature Profile

Figure 18 shows the vapor and liquid flow rate for each section. In contrast to the conventional single vertical column, the plates of the enriching section and of the stripping section are mounted side by side in parallel. As the vapor rises in both sections, the vapor rate in the enriching section decreases, while that in the stripping section increases.

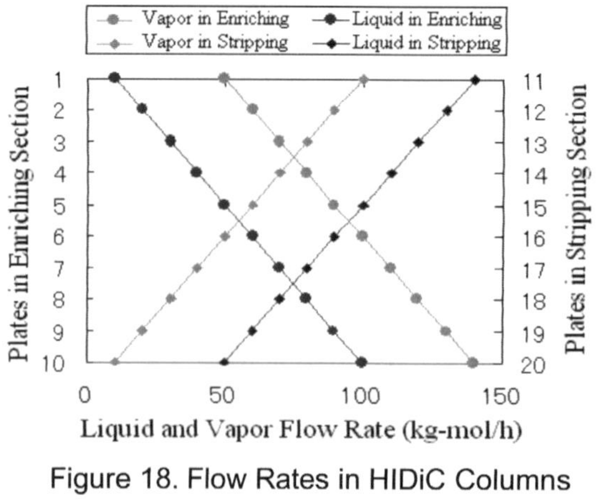

Figure 18. Flow Rates in HIDiC Columns

Figure 19 compares the flow rate of the HIDiC to that of a conventional system. In the conventional system the flow rate is constant, whereas in the HIDiC, the flow rate changes. The figure also shows the HIDiC's "average flow rate." From this flow rate, we calculated the reflux ratio, which was 0.9. This is less than the minimum reflux ratio.

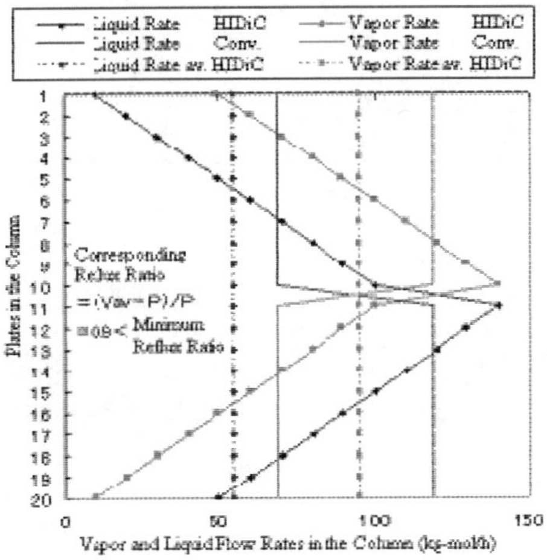

Figure 19. Flow Rates in HIDiC and Conventional Columns

5. CONCLUSIONS

We have presented the ground-breaking design of HIDiC and efficient operation of the HIDiC pilot plant. We have demonstrated why HIDiC has such a high capability for saving energy — over 60% greater than that of conventional systems. HIDiC also offers some unusual and innovative features. Among them is the fact that HIDiC does not require external reflux, because it can produce internal reflux through internal heat exchange. Moreover, the internal reflux ratio with HIDiC is less than the conventional minimum reflux ratio.

Acknowledgments

We gratefully acknowledge the following persons, agencies and companies.

HIDiC Project: Dr. M. Nakaiwa, Project Leader

HIDiC Review Committee: Prof. Y. Naka, Prof. S. Koyama; Mr. Y. Saito, Mr. K. Matsuda, Mr. S. Midori

Government representatives of the Ministry of Economics, Trade and Industry (METI) and New-Energy Development Organization (NEDO)

Companies: Maruzen Petrochemical Co. (pilot plant site host), Kimura Chemical Equipment Co., Kansai Chemical Equipment Co., Taiyo Nissan Co., Kobe Steel Co.

Unfixed Wall: the Key to a Breakthrough in Dividing Wall Column Technology

Bjoern Kaibel, Julius Montz GmbH, Hofstrasse 82, Hilden, 40723, Germany

Employing unfixed dividing wall proved to be the key to real breakthrough of dividing wall concept in the field of packed distillation columns. The flexibility regarding the positioning of easy assembling partition walls along the column extended the range of applications with respect to composition, relative volatilities and thermodynamic condition of three component mixture feeds and made practically feasible distilling four or more component mixtures in one shell. Similar benefits are expected from this highly flexible technology if applied in conjunction with trays, which is a more demanding task regarding the type and the size of potential applications. However, in addition to process related gains there are also numerous practical benefits of implementing unfixed wall concept in the field of tray distillation columns. It allows easy revamps of existing columns. Simple construction reduces costs significantly as well as the installation time. Welding work is reduced to an absolute minimum, i.e. there are no mechanical stresses introduced in the column shell through thermal expansion. No extra codes have to be followed regarding the tolerance requirements of the dividing wall. These and other, process and mechanical design/operation related benefits will be demonstrated using practical examples.

Inside-Out Approach for Optimizing Distillation Design

By

Darius Remesat, Koch-Glitsch Canada LP

Presented in the Distillation Topical Conference
of the
2007 Spring AIChE Meeting,
Houston, TX,
April 22-26, 2007"

Previously published in Hydrocarbon Processing (HP) August, 2006

Inside-Out Approach for Optimizing Distillation Design

By

Darius Remesat, Koch-Glitsch Canada LP

The inside-out (integrated) design approach has been advocated in many process design presentations, papers, and books[1]. This article provides a compilation of recent successful case studies from various processing areas, such as refining/upgrading, high purity petrochemical, and air separation, that demonstrate the economic value to designers and operators in having heat exchange and tower internals experts involved in the initial development of a process design.

Inside-Out Approach

The basis for the inside-out approach, as the name implies, is to start the design at the heart of the separation process, the distillation column, and work outwards to the supporting equipment. The key difference from the more traditional sequential approach, as the comparison in Figure 1 illustrates, is that the column sizing/design is moved up in the process and a recycle step involving equipment experts is included. The "out" in inside-out also stands for reaching out to the equipment vendor experts to gain from their vast experience and exposure to many situations. Equipment expert input can provide significant value as demonstrated in the following examples.

Many designers argue that the column design is an early design milestone and that appreciable recycle is provided in designs; however, with the reality of tight schedules and rushes to book long-lead items (i.e. vessels), the true nature of the inside-out approach is rarely followed. Because tower internals and heat exchangers comprise a small part of a project's cost and typically are not critical path equipment items, column and heat exchanger sizing are placed at the end of the design queue. As a result, there is very little room for any optimization because all the equipment expert has to work with is meeting a required specified duty to design an exchanger or using specified liquid and vapor flow rates to design the tower internals in a set diameter. The notion of a proven or standard design and safety margin is the justification for not embracing the experience from a third party equipment expert.

The significant process and mechanical benefits captured from an inside-out approach far outweigh the additional investment in time.

Process benefits come from capturing energy efficiencies and/or increasing product yield through:

- Embedding detailed equipment characteristics into process simulation vs. typical rules of thumb
- Improving flowsheet topology
- Optimizing column staging

Mechanical benefits come from incorporating the appropriate considerations by

- Using specifications developed from operating, construction, and shutdown experience
- Using a modularized approach to simplify field work

These process and mechanical benefits result by following the inside-out approach detailed in Figure 2. Once the initial economic objectives, cycle screening, and simulation are developed, the steps highlighted in green use the detailed input from an equipment expert (ideally an experienced equipment vendor expert) to build the equipment scope from the inside (distillation process) out. The process evaluation continues to ensure no equipment envelopes are violated when necessary iterations are performed.

One of the keys to making this approach successful is finding an available in-house expert or a vendor with a long-term, life-cycle approach that is willing to provide the necessary scoping support. The following examples indicate the search for an equipment expert to assist in the initial design phase is well worth the effort.

Inside-Out Design Case Studies

1. A&V Unit – Grassroots Design

Examples 1 and 2 involve new designs for an atmospheric and vacuum unit to process heavy synthetic Canadian crude. In particular for the vacuum column, significant capital and operating value was obtained from fully incorporating the performance benefits of the tower internals and supporting equipment into the process design (via simulation). A few examples of the equipment that can impact the way a process design develops are:

Feed inlet device (Radial enhanced vapour horn[2]) – maximizes open area, reduces entrainment potential

Grid/packed combo wash bed section – mitigates fouling, optimizes heat transfer

Spray distributors – mitigates fouling, provides ample wetting coverage

Integrated heater and transfer line – improves plot plan layout, simplifies piping, provides tower diameter reduction

In both examples, Table 1 shows that the optimized design reduced (or "right sized") the column diameter and provided increased production upside while maintaining all other dimensions and satisfying the base capacity and yield requirements. Both designs used the same simulation topology[3], but the optimized case had an iterative step, noted in Figures 1 and 2, that factored in the performance of the internals equipment with respect to the crude characteristics and performance specifications.

As an example, the enhanced radial vapor horn allows increased lift velocities for a given entrainment rate. Figure 3 provides a sketch of the vapor horn on the left with CFD output[4] showing superior, field-validated vapor distribution. Knowing the performance the device is capable of gives the process designer an opportunity to refine the design and optimize the associated equipment in the column as well.

Figure 4 illustrates the relative benefit, compiled from numerous operations, of using an enhanced radial vapor horn versus a standard inlet feed device. The ability to reach a higher C-factor for an acceptable level of entrainment provides a design with a greater ultimate capacity[5]. This use of an equipment's specific features needs to be done in the initial stages of the design and may require some iteration to finalize a suitable design.

Without timely expert input, a sub-optimal design is created because a key piece of data is not factored into the design. A more conservative design could result in greater capital and operating costs. It could also provide future revamp opportunities, which is good news for the equipment vendor. However, the economic impact of not optimally sizing the tower from the beginning should be fully evaluated.

Test run data for example 1 reveals that the balanced and optimized design has the column running at a continuous production of 32,000 BPD. As a function of the features of the inlet vapour feed device and associated equipment, the desired product quality is maintained at very high C-factors of 0.46-0.47 ft/s in the flash zone. C-factor is defined as

$$C_f = v_s \sqrt{\frac{\rho_v}{\rho_l - \rho_v}}$$

where v_s is the superficial velocity

ρ_v is the vapor density

ρ_l is the liquid density

For example 1, the approximate 5-year Net Present Value (NPV, 15% internal rate of return) is approximately $38 MM Canadian (CAD) based on:

- Actual capital savings from a reduced column diameter (~$20 MM in equipment/construction costs)
- Increased operating revenues from taking advantage of the increased capacity potential (For example: $5.3MM/yr from 1400 BPD @ $8/bbl upgrade)

These benefits would not have been possible if the equipment options were plugged in after the design had been set. Whether there are real benefits to be obtained can only be answered if the inside-out analysis is performed.

2. Vacuum Column – Phased-production Requirements Design

Adding to the vacuum column design optimization from the previous case, the following case study incorporates a phased feed-supply situation. Many oil sands upgrader operators implement phased production upgrades to match the output profile from bitumen deposits. As part of the vacuum column design process, an evaluation was performed to determine if a vacuum unit could be sized for ultimate production while handling the initial lower feed rates. The base option was to have multiple smaller units constructed in phases to match the feed supply profile.

To provide the best possible scoping answer for a valid comparison with the base option, external equipment experts were brought in at the beginning of the design process to determine the possibilities for using an ultimate-capacity column. Figure 5 shows the configuration of the entire vacuum column with the portion of the column that included features to allow a phased approach to match the feed supply.

Specific mechanical and process considerations were provided in the design to ensure reliable operation (eg. prevent coking), mechanical integrity, and a reasonable transition plan. Table 2 shows the difference in operation between phases 1 and 2. The cost of reducing the effective cross-sectional area in phase 1 was compared to the savings from reducing the pumparound flow rates in a staged design. The staged design incorporated the planned internals modifications in preparation for a step change in production. The work during a planned shutdown was minimized, accomplished on time, without an incident, and the subsequent startup was smooth. During the early phases, unnecessary energy was not used because appropriate pumparound rates matched the column feed, not the physical dimensions of column.

The Net Present Value (NPV) calculations incorporated the cost of one unit versus two, the need to shut down to make modifications, and the operating costs for the two units. Many iterations and detailed calculations were needed to ensure the right value to the benefits were included. Without the collaboration with equipment experts, the one-unit staged design would not have been realized. The NPV from the one large column approach was significant enough to make a decision in favor of proceeding with the single vacuum unit approach.

Even if multiple units are chosen over one large unit, which is becoming more prevalent because of the ability to better manage the project scope, there is opportunity to improve these more standard designs by using the design expertise from equipment experts, as noted in the first two examples.

C_3 (Propane/Propylene) Splitter – Grassroots Design

For C_2 (ethane/ethylene) and C_3 splitters, the inside-out design approach is used extensively; however, relying on a specific equipment expert throughout the process to provide performance characteristics for the design may not provide the optimal solution. This case study highlights the need for the designer to address this issue. The example does show value from engaging multiple equipment experts to produce a suite of solutions for evaluation purposes.

Different internals exhibit different performance characteristics (efficiency and capacity) with the potential for different tower solutions. On a standard crossflow tray at the vapor liquid interface, vapor travels upward interacting with liquid travelling in the horizontal direction. A counterflow tray has the liquid essentially travelling downward with the vapor travelling upward at the interface. A standard crossflow tray will typically provide greater efficiency and less capacity than a standard counterflow tray. The performance comparison between enhanced cross flow and enhanced counter flow is tougher to categorize. As a result, the process designer in this situation should provide a generic enough process scope to allow multiple third-party equipment experts to evaluate. The operator/designer will benefit from the choice of multiple "local" optimum solutions provided by each equipment expert when the designer does the following:

- Uses the inside-out approach (Figure 2).
- After the design basis step, includes various equipment experts to size the column and optimize the staging (noted in green in Figure 2).

The iterative process will then provide the best choice given the original design basis.

As Table 3 illustrates, two different equipment experts should come up with different solutions due to the efficiency and capacity differences between specific equipment types. In this case, the biggest difference is the choice between one column and two columns with the obvious reduction in the project cost for a single column.

During the iterative step in the inside-out approach, other variables, such as pressure, reflux ratio, and enhanced distillation trays, can be manipulated to meet the initial operating design specifications. A trade-off analysis between project cost and operating cost will result. The designer benefits from the choice of multiple feasible solutions that otherwise would not be possible during the downstream bid phase. If solution A was only considered during the initial design phase, the opportunity to provide a solution B during the equipment bid phase is greatly reduced because the vessel design parameters have been set. The designer increases the opportunity for an optimal solution by allowing equipment experts to apply unique equipment performance characteristics during the column sizing/stage count optimization step.

3. Air Separation Unit – Re-Design

Power costs are the primary cost input in the separation of air. In response to an increase in these costs, a standard ASU design (Figure 6) was evaluated to determine the additional capital that could be justified to meet a goal of 10-12% operating efficiency improvement of the plant. Numerous new enhancements (Table 4) could be included in the new design.

A typical method in the design improvement process is to simply "plug" the new technologies into the existing process design. The intent is to make the new technology fit the process and not the process fit the technology. This sequential method saves engineering time and can be easily input into a schedule with distinct milestones; however, it does not capture the full value from the enhancements developed into the equipment. This example shows the benefits of the inside-out approach by starting with the enhanced equipment (inside), capturing its efficiency benefits, and then building the simulation around (outside) the entire plant.

Table 4 shows the efficiency gains from implementing the various enhancements for each design approach
for a 600 TPD air separation unit with multiple products (liquid oxygen, gaseous oxygen, liquid nitrogen, and liquid argon).

For the same product flow, the integrated design approach provided a 50% increase in efficiency over the sequential approach (Table 4). Keeping the column diameters constant from the base case, the inside-out design approach provided a maximum efficiency improvement of 13% coupled with an unexpected, yet welcomed, higher capacity (~15% more production). Figure 7 shows the performance comparison from the two design approaches relative to oxygen gas production. Assuming typical capital costs ($17 MM CAD base install, 15% IRR) and operating costs ($.05/kwh power), the added value from using the inside-out design approach is over $1.0 MM/yr per plant. The additional time/cost to perform the inside-out design is small compared to the value gained from an optimized design that can be replicated many times.

In the inside-out approach, the lower pressure drop provided by the new next-generation heat exchanger technology can be fully realised by incorporating this feature into the compression design and mass transfer stage count. In a sequential approach, the duties are provided as a target, and the heat exchanger expert designs to meet that target. As a result, the sequential approach provides a 3% efficiency gain, which is an improvement although it does not capture the full value from the new equipment design features as demonstrated by the inside-out approach (4-5% efficiency gain). The full value provided by the heat exchanger enhancement is captured in the inside-out approach.

Conclusions:

The inside-out approach provides a successful methodology for a designer to follow that provides an optimized design by incorporating appropriate equipment characteristics into the design. The examples in this paper reinforce the value of the inside-out approach and highlight the value of engaging external equipment experts in the early stages of the design process. In addition, the case studies illustrate that many processes, from heavy oil upgrading to polymer feed conditioning, can benefit from the inside-out approach.

Figures:

Figure 1 – Sequential vs. Inside-out process approach

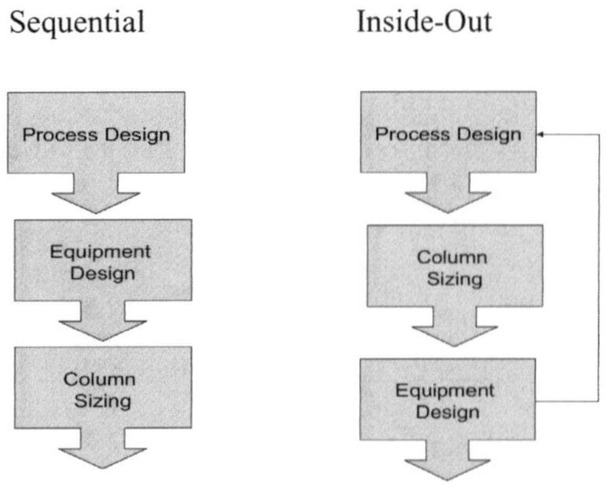

Figure 2 – Inside-Out Approach Work Process Flowchart

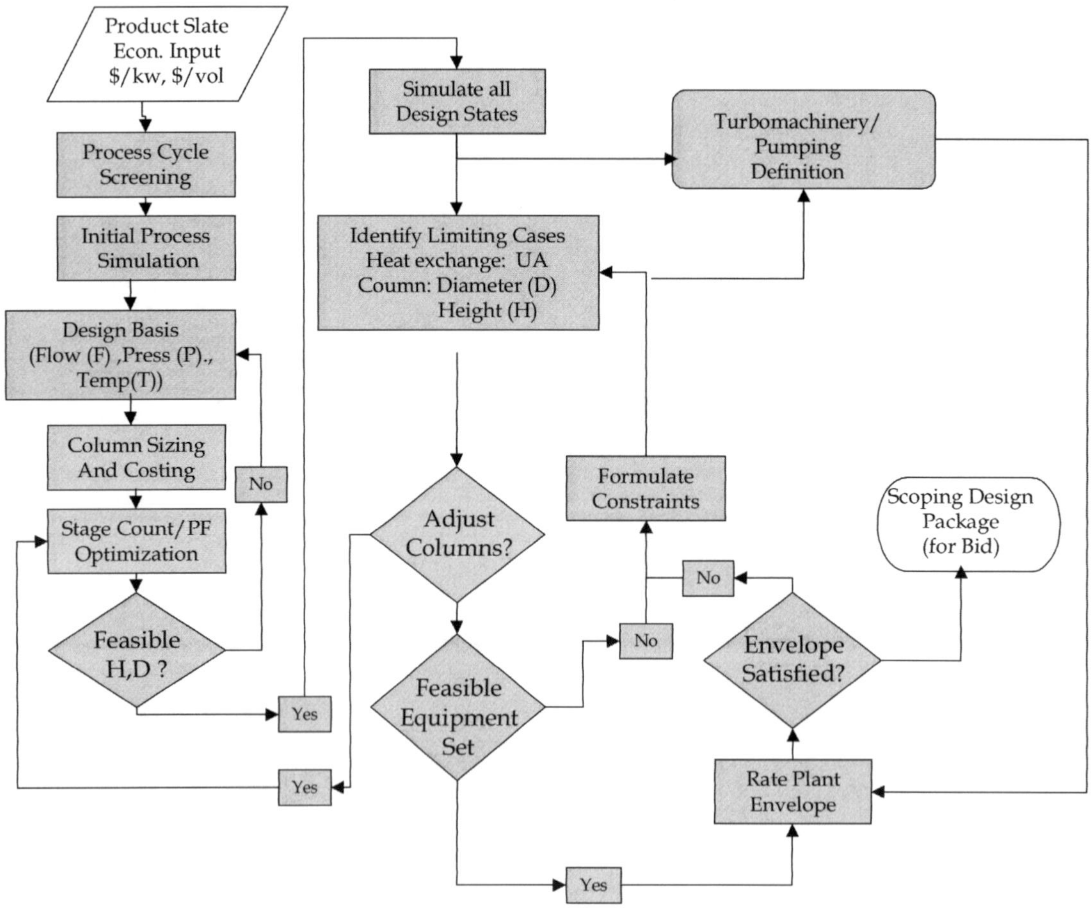

Figure 3- Vapor Horn Sketch with CFD output for flow distribution[5]

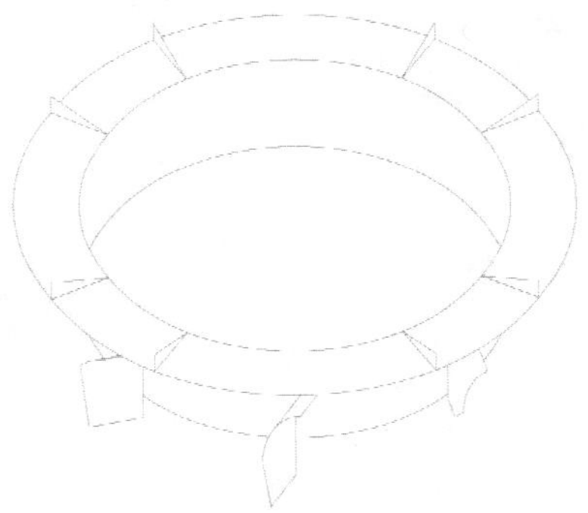

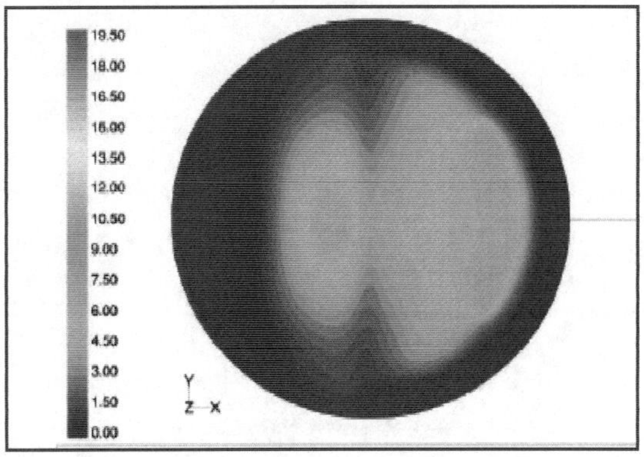

Figure 4- Inlet Feed Device Relative Performance comparison

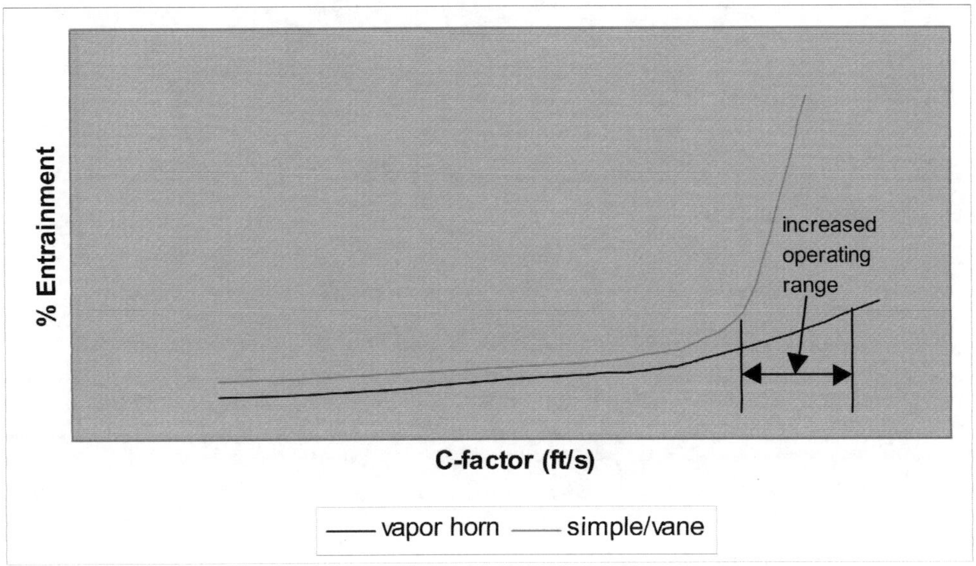

Figure 5 – Vacuum Column – Phased Approach

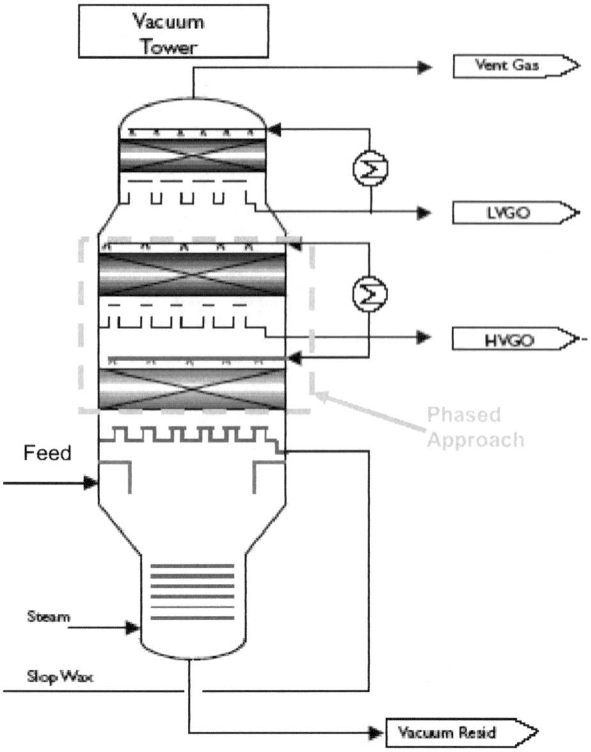

Figure 6 – Sketch of a Typical ASU Unit

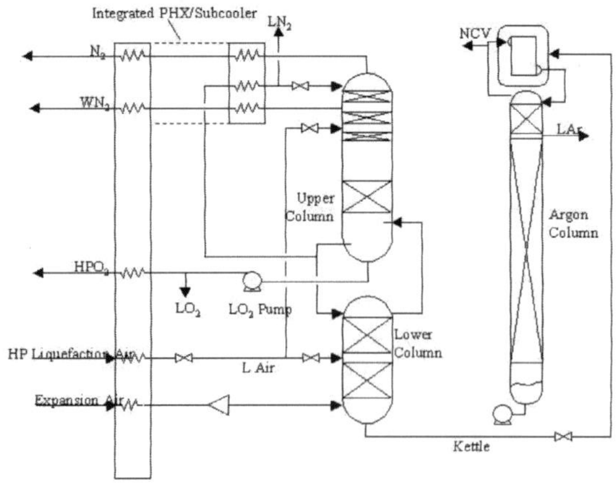

Figure 7 – Performance differences between Inside-Out and Standard Design Approach

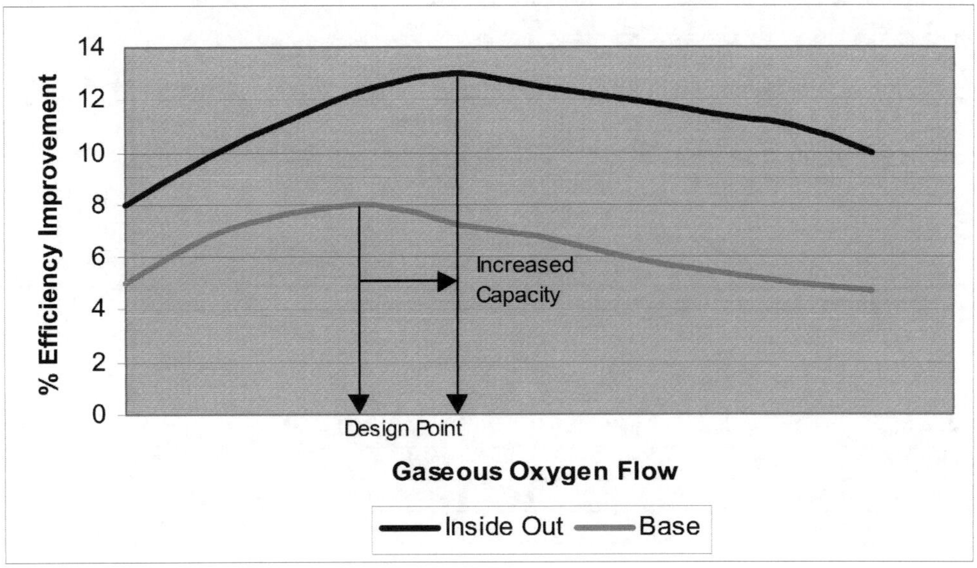

Table 1- Comparison between the base design and the design provided by the integrated design approach.

	Example 1		Example 2	
	Base Case	Integrated Case	Base Case	Integrated Case
Crude (API)	20.1		10.5	
Base Feed (BPD)	25,000		250,000	
Design C-factor (ft/s)	0.35		0.35	
Above Flash P drop (mmHg)	13		15	
Operational mode	Damp, no strip steam		Damp, no strip steam	
Design Margin	20%	25%	15%	20%
Diameter	18'	16'	44.5'	41'

Table 2 – Phased Vacuum column design approx.

	Phase 1	**Phase 2**
Feed (kBPD) – normalized to 100 for phase 1	**100**	**160**
Crude Average API	**7.2**	
Effective Tower X-section area %	**74**	**100**
Wash Oil Pumparound (BPD) – normalized to 10,000 for Phase 1	**10,000**	**14,000**

Table 3 – Internals options for C3 splitter grass roots design.

Item	Solution A	Solution B
Feed ,kgmol/hr/%propylene	1100/48	
Ovhd Pressure	175 psia	
C3 product, type/spec	Polymer grad/ 99.5%	
Operation	Heat pump, optional pasteurization	
Diameter	16'	
# of columns	1	2
Type of Standard tray	Cross flow	Counter flow
# of actual trays	213	272

Table 4 – ASU efficiency improvements – Sequential vs Inside-out design.

Item	Increase in Efficiency	
	Sequential	Inside-Out
Trays to High performance Packing	3%	4%
Pre-purifier enhancements	0%	1%
Compression/Booster enhancements	2%	3%
Next Generation heat exchangers	3%	4%
TOTAL	8%	**12%**

References:

1 Dimian, A., C., "INTEGRATED DESIGN AND SIMULATION OF CHEMICAL PROCESSES", Volume 13, Elsevier Science, May 2003

2 Ze**idan, A, "Consider a practical approach to vacuum unit revamps" Hydrocarbon Processing,**

January 2006

3. Golden S., Shah, V., and Kovach III, J., "Improved Flow Sheet Topology for Petroleum Refinery Crude Vacuum Distillation Simulation ", Presentation, 44th Canadian Chemical Engineering Conference, Calgary, 1994

4 Koch-Glitsch LP, "Intalox Packed Tower Systems, Severe Services", Bulletin KGSS-1, 2003, Koch-Gltisch LP

Guidelines for BTX Distillation Revamps

Karl Kolmetz[1], Joseph C. Gentry[1], C. S. Kumar[1], and Jeff N. Gray[2]. (1) GTC Technology LP, Singapore, Singapore, (2) KLM Technology Group, #03-12 Block Aronia, Jalan Sri Perkasa 2, Taman Tampoi Utama, 81200 Johor Bahru, Taman, Malaysia

There are many reasons to revamp a process unit. These include increased purities, increased recoveries, decreased environmental impacts and increased capacity. The best revamps will include each of the listed items.

With the current environmental mandates to reduce the benzene in the gasoline pool, the production of benzene for use in other products is increasing. In the United States the benzene in the gasoline pool is regulated to very low levels, whereas in other parts of the world the standard is 1.0 weight percent. As this world standard is lowered this benzene will need to be purified for downstream reprocessing.

Currently many of the benzene separation units are utilizing liquid / liquid extraction which can be revamped to extractive distillation for increased purities, increased recoveries, decreased environmental impacts and increased capacity. The authors will examine the choices for revamping benzene distillation, review a case study of an April 2006 revamped liquid / liquid extraction to extractive distillation, and guidelines for future revamps.

Liquid Continuous Distillation

Frank Seibert[1], J. Christopher Lewis[2], and James Fair[1]. (1) Separations Research Program, University of Texas at Austin, 10100 Burnet Road, Bldg 133/CEER, Austin, TX 78758, (2) Separations Research Program, UT-Austin, 10100 Burnet Road, Bldg 133/CEER, Austin, TX 78758

The bubble column is a simple arrangement of a vessel containing a gas sparger and little else. When operated on a continuous basis, liquid is fed to the column at some selected elevation, and gas is passed up through the vessel in the form of bubble swarms. The device has been used extensively for gas-liquid reactions. Variations include the use of suspended solid catalysts and slurry feeds, in both cases taking on a more complex character because of the third phase. To deal with heat effects, cooling surfaces can be installed within the vessel. There have been some cases where trays have been inserted to mitigate the strong trend toward complete mixing of liquid in the contactor. Some example process applications are fermentations, hydrogenations and oxidations. Experience with bubble columns has been limited to very low gas superficial velocities (< 0.03 m/s), with equivalent fractional gas holdup values of 0.1-0.3, and with unidirectional mass transfer. A large literature exists on bubble column operating characteristics operated under non-distillation conditions at low loadings. In this paper we go far beyond the usual bubble column operating ranges, and report data on liquid-continuous operation up to the capacity limitations of the usual vapor-continuous distillations using trays and packings. The work has been done in the distillation system pilot plant of the Separations Research Program (SRP), with a column diameter of 0.51 m and use of a conventional overhead condenser and a kettle reboiler. Comparisons are made with conventional trays and structured packings for pressure drop, gas holdup, and mass transfer efficiency.

New Energy-Efficient and Low-Cost Multicomponent Distillation Configurations

Arun V. Giridhar, Rakesh Agrawal, and Venkat Venkatasubramanian. Purdue University, 480 Stadium Mall Drive, West Lafayette, IN 47907

The synthesis of efficient multicomponent separation networks has posed computational challenges in process design, with distillation networks having the greatest economic impact. One reason for the computational difficulty is that there are a combinatorially large number of possible configurations for a given multicomponent feed. Most configurations designed and built in the last 40 years have occupied a very small fraction of this large search space. Recent work in the literature has shown the energy efficiency of some new configurations, which were developed through a process of discovery, which have not been included in traditional sets of configurations. It is desirable to build a systematic framework that embeds a "complete" set of configurations, so that advances in energy efficiency are not reliant on discovery processes. In this work, we introduce a novel framework to embed all possible distillation configurations for a given feed subject to the constraint that they use the minimum number of columns (i.e., n-1 columns for an n-component feed). We believe that this framework and the associated method are powerful enough to provide energy-efficient and cost-effective designs of distillation networks when used on a computer, and simple enough to enable humans to verify the quality of the design by hand in reasonable time. We accomplish this by introducing the concept of a "supernetwork", which considers only configurations with the minimum number of separation units, both with and without thermal linking. Several of these configurations have not been previously described in the literature. Our detailed comparison of the heat demand of such configurations with traditional sharp-split configurations for dozens of feed configurations has shown savings potential of up to 45%. We also evaluate different optimization methods to search the novel configuration space and optimize the heat demand with respect to composition and flow rates.

Energy Saving by Integration of Refining and Petrochemical Plant

Tadao Matsumoto, Petroleum Refining Technology Center, Idemitsu Kosan Co., Ltd., 26, Anesaki-Kaigan, Ichihara City, Chiba, 299-0107, Japan

The catalytic naphtha reforming unit produced high octane product for gasoline and aromatics for petrochemical products in Tokuyama refinery of Idemitsu Kosan, Japan. An half amount of the high octane product were fed to four distillation trains to produce mixed xylene and gasoline pool. The mixed xyene and remaining high octane product were the feedstock of a para-xylene unit in petrochemical facility which located neighbored the refinery.

We investigated the possibility of saving energy by integration of the four distillation trains and the para-xylene unit by using a process simulator. We examined how to maintain para-xylene and gasoline pool production and our solution was that only changing the operating conditions of the first and second distillation column, and the third and forth distillation columns do not need to operate to produce mixed-xylene in the refinery. The new operating conditions were that C6- fraction was distilled in the first column and raw toluene was separated as distillate in the second column. The bottom of the second column, the mixture of mixed-xylene and C9 aromatics heavier would be the feedstock of the para-xylene unit. As the feed-rate of C9 aromatics column in the para-xylene unit became approximately twice of the original design, the trayed column should replaced with a new forth generation random packing, Raschig Super-Ring.

We carried out successfully energy conservation job that needed some modifications of the piping, storage tanks, installation of a new heat- exchanger for raw toluene and replacement of the internal of C9 aromatics column during turnaround in 2004. The two distillation columns in the refinery and the para-xylene unit are running well after the turnaround. Energy saving which equivalent to 6,000 KL of fuel oil was achieved by the integration of refining and petrochemical plant

Separation of P-Diethylbenzene from the Binary Diethylbenzene Mixture by Distillative Freezing

Lie-Ding Shiau and Keng-Fu Liu. Chemical and Materials Engineering, Chang Gung University, 259 Wen-Hwa 1st Road, Taoyuan, 333, Taiwan

A new separation technique, called distillative freezing (DF), has been successfully applied to separate p-xylene from a liquid mixture of m-xylene and p-xylene (Shiau et al., 2005). Basically, DF is a distillative crystallization technology, which combines distillation and crystallization to result in pure crystals. The DF process is operated at triple point condition, in which the liquid mixture is simultaneously vaporized and solidified due to the three-phase equilibrium. It results in the formation of pure solid, and liquid phase and vapor phase of mixtures. The process is continued until the liquid phase is completely eliminated and only the pure solid crystals remain in the feed. p-diethylbenzene (p-DEB) is used as a crosslinking agent in the synthesis of polymeric resins and the desorbent in the Parex process for the separation of p-xylene from the mixed xylenes. To produce pure p-DEB, separation of p-DEB from the mixed DEB isomers is encountered in the production of p-DEB by the ethylbenzene alkylation. Due to close boiling temperatures of the mixed DEB isomers, it is rather complicated to separate them by distillation. The objective of this work is to investigate the feasibility of DF in separating the mixture of p-diethylbenzene (p-DEB) and m-diethylbenzene (m-DEB). The experiments show that p-DEB can be purified by DF from 95% to 99% or even from 90% to 99%. The experimental recovery rates are in the range 58~77%.

Shiau, L. D.; Wen, C. C.; Lin. B. S.,Separation and purification of p-xylene from the mixture of m-xylene and p-xylene by distillative freezing"Ind. Eng. Chem. Res. 2005, 44, P2258-2265.

Crude Unit Overhead Corrosion at Sincor

German A. Luna Sr., TECHNICAL SERVICES, SINCOR C.A., EDIF ADMINISTRATIVO OFICINA PB-7, COMPLEJO INDUSTRIAL JOSE, BARCELONA, Venezuela

The SINCOR Upgrader crude oil column had shown several corrosion problems since startup early 2002. Design conditions did not considered the risk of having free water inside the column neither salt formation inside the tower, this was not considered to be a problem until after 10 months of operation, overhead condensers started to fail. Preliminary investigations indicated that at operating conditions and enough NH3 levels the salt formation inside the column was something to be expected to happen. The complexity of this system is due to the presence of NH3, HCl, H2S, and Light Organic Acids (LOA) that competed amongst them.

Due to Heavy Crude characteristics, LOA formation well above 200 ppm is expected when running the outlet of the heaters around 374 C (design condition) and was not considered in design either. Some other tests indicated presence of NH3 in the formation water in the crude, and evolution of H2S in more than 50 ppm in the Overhead water.

New operating conditions were set at 135 C (min top temperature), top P/A return temperature of min 127 C, and continuous wash water injection to keep overhead condensers clean. The column will be monitored and gamma-scanned frequently until next T/A to follow up the packing conditions when operating at these conditions.

Tower Internals Modifications Improve Performance of Atmospheric Crude Tower

Pat Quotson, Koch-Glitsch, LP, 4111 E 37th Street N, Wichita, KS 67220

ENAP Refinerias operates a crude unit topping column at the Bio Bio refinery in Talcahuano, Chile. The refinery engineers expressed concern with discoloration of the atmospheric gas oil (AGO) product, which could be symptomatic of entrainment of heavy reduced crude into the AGO draw collector. Plugging of the liquid distributor above the stripping section was observed during previous shut-downs and this was considered to be a major contributor to the poor AGO product quality. Koch-Glitsch performed a detailed hydraulic analysis of the existing equipment below the AGO draw collector using process simulations of the current topping unit operation. Based on this analysis, the wash tray weirs were modified, down-pipes were added below the bottom wash tray seal pan and the stripping section liquid distributor was replaced. The new equipment was designed, fabricated and installed within a three week period. After the column was put back into service the atmospheric gas oil color improved and its metals content decreased significantly. This paper describes the details of this revamp.

A CFD Study Comparing Different Feed Nozzle Arrangement within an Empty Spray Section in a Coker Fractionator

Karolline Ropelato[1], **Leonardo Rangel**[1], Edson R. Marins[2], and Washington de O. Geraldelli[2]. (1) ESSS - Engineering Simulation and Scientific Software, CELTA - Rod SC-401, Km 01, Florianopolis, 88030-000, Brazil, (2) CENPES (R&D Center), PETROBRAS, Cidade Universitaria, Q7, Ilha do Fundao, Rio de Janeiro, 21941-598, Brazil

Recent study was developed to evaluate the vapor distribution and its influence on the heat and mass transfer process considering coker fractionator designs with only one feed inlet nozzle. In these cases, high vertical velocities can be observed near the wall opposite to the inlet nozzle and large recirculation on the flash and washing zone, causing less efficiency on the heat and mass transfer between liquid and vapor phases. The vapor maldistribution on this section must be avoided since this behavior increases the probability of undesirable coke formation. The main purpose of this study is to predict the fluid dynamic behavior in a coker fractionator washing zone and analyze the influence of vapor distribution on the heat and mass transfer efficiency, considering different number of feed inlet nozzle. In general, licensors are used to design coker fractionator with only one feed inlet nozzle. However, recently, in new coker designs, specially those with four coke drums, it has been used a different design concept on the transfer line from the coke drum to the fractionator and two feed inlet nozzles has been used in these cases. For the coke drum process design, the most important criteria is the maximum vapor velocity and the number of coke drums are determined by the unit capacity and the maximum diameter limit that it can be manufactured. Depending on the unit capacity, six or more coke drums may be necessary. Six coke drums unit has become quite common recently and a new configuration of transfer lines from the coke drums to the fractionator and multiple inlet nozzles can be developed. A study comparing different number of inlet feed device was carried out in a coker fractionator washing zone considering a empty spray section instead of a conventional liquid-vapor contact device as tray and/or packing. Equipment with empty spray section takes some advantage such as lower: pressure drop, height, total costs and tendency of coke formation in high temperature sections. The results of this work are based on a steady state, three-dimensional, gas-liquid model with an Eulerian-Lagrangian approach of spray distributors using computational fluid dynamics techniques (CFD). A non-ideal Soave-Redlich-Kwong equation of state was applied to model the liquid phase equilibrium and fluid properties calculation within the condensation and evaporation models. The CFD model was capable to predict the fluid dynamic behavior of multiple inlet nozzles in a coker fractionator, and has proved to be a powerful tool to help engineering decisions. KEYWORDS: Coker fractionator, washing zone, empty spray section, multiphase flow, CFD.

Systematic Field Tests Turn Loss into Gain in High-Capacity Trays

Part 1: Excursion

By

Henry Z. Kister, Fluor, Aliso Viejo, CA, USA

and

Neasan O'Shea (retired) and Dan Cronin, ConocoPhillips Ireland Limited, Whitegate Refinery, County Cork, Ireland

Presented at the Distillation Topical Conference, AIChE Spring Meeting, Houston, Texas, April, 2007

Summary

This paper originates from a systematic troubleshooting effort that successfully diagnosed two complex independent problems which severely limited the capacity of an atmospheric crude distillation tower. The crude tower had been retrayed (by others) with modern high-capacity trays, but upon start-up achieved less than 80% of its previous capacity.

The troubleshooting identified two bottlenecks: premature flooding on some of the heavy naphtha-diesel fractionation trays, and a phenomenon which we termed "excursion," in which sudden temperature reductions around the diesel draw, accompanied by distillate yield loss, occurred upon rate increase.

This paper focuses on the excursion bottleneck. Systematic field tests showed conclusively that the excursions occurred when the diesel draw rate exceeded about 2000 tons/day (t/d). This was lower than the diesel flow rate that could be drawn prior to the retray. The new bottleneck was diagnosed to result from a seemingly benign modification to the downcomer that transported diesel to its draw pan. This modification halved the degassing time in the downcomer, leading to aerated liquid reaching the diesel draw pan. The excursions occurred when the diesel draw nozzle and rundown line reached their self venting flow limit. The excursion reflected the switch from self-venting flow to choked flow in the diesel rundown line.

Based on this diagnosis, some of the diesel was drawn out of an available kerosene draw nozzle further up in the tower. This unloaded the diesel draw, keeping it in the self-venting mode. This mode of operation also helped the premature flooding, which was the second bottleneck in the tower (this will be described in another paper). As a result, drawing kerosene permitted the tower to recover almost all the capacity loss without any hardware modification. The longer term solution was to replace the diesel draw by a seal-welded chimney tray and to increase the sizes of the diesel draw nozzle and rundown line. These modifications completely eliminated the excursions, permitting higher-than-design capacities to be achieved.

Systematic Field Tests Turn Loss Into Gain in High-Capacity Trays

Part 1: Excursion

Henry Z. Kister, Fluor, Aliso Viejo, CA, USA and Neasan O'Shea (retired) and Dan Cronin, ConocoPhillips Ireland Limited, Whitegate Refinery, County Cork, Ireland

Introduction

The last 2-3 decades saw tremendous improvements in high-capacity trays and packings. Modern trays and packings have permitted towers to be debottlenecked to throughputs and separations not previously envisioned. Yet, occasionally a high-capacity tray or packing debottleneck falls short of achieving its design targets.

A common reaction is to blame the trays or packing, and to seek a sledge-hammer solution, such as abandoning the new trays and replacing them by structured packing, or even building a new tower. Sledge-hammer solutions like this are not only costly, but are also highly counter-productive. Often they may obscure simple solutions that can be implemented immediately (rather than at the next turnaround), and minimize losses and down time.

In most situations where high-capacity trays or packings fail to achieve their design expectations, it is not that the trays or packings are "no good." Rather, it is hydraulic details associated with the trays or other internals that cause the premature bottlenecks. The correct strategy to address these is not tossing the trays out; rather, it is good troubleshooting that seeks the root cause and eliminates the real bottleneck.

Good troubleshooting is systematic and thorough. It combines interacting with the operating staff to collect operating experience and obtain a good problem definition, conducting well-targeted field tests, applying process diagnostics (such as gamma scans and surface temperature surveys), simulation, and hydraulic analysis. A vital element is good testing of each theory and hypothesis with valid field tests. One good field test can humble a thousand expert opinions. This approach was adopted by our team and will be demonstrated in this paper.

This paper describes a case history where a high-capacity retray of a crude fractionator (by others) led to a capacity loss. A thorough and systematic investigation utilizing extensive field tests identified and diagnosed the causes of two completely independent bottlenecks that caused the revamp to fall short of its expectations. The diagnosis led to a very simple on-line solution that overcame both bottlenecks, involved no hardware changes, cost little to implement, and recovered most of the losses almost immediately, circumventing any need to shut the crude unit down. Longer term, the diagnosis led to correct, low-cost internals modifications that produced capacities well-above the design.

Background

In the 1990's, Irish Refining Company operated a 52,500 BPSD (7,070 t/d) crude unit processing North Sea crudes (This refinery is now owned by ConocoPhillips Ireland Limited). Products from the unit included light and heavy naphthas, diesel, and atmospheric residue. The diesel was produced to a cloud point spec of 0°C while the heavy naphtha (HN) was produced to an ASTM D86 95% point spec of 200°C.

Figure 1 shows the unit with the mass balance data of August 23, 1993, with the unit operating at a reduced crude rate of about 38,000 BPSD (5,200 t/d).

The atmospheric tower contained two pumparounds (BPA and TPA) and two active side draws. The upper side draw was heavy naphtha (HN), the lower was diesel. The diesel draw was stripped in a side stripper.

Prior to 1993, the 4 meter ID atmospheric crude tower contained 34 single-pass jet tab trays. In May 1993, the tower was revamped to increase fractionation efficiency and to permit the refinery to intermittently take a kerosene side-cut while maintaining the same throughput of 52,500 BPSD (7,070 t/d). To achieve this, trays 11 – 20, i.e., the trays in the HN-diesel fractionation section, were replaced by single-pass high-capacity trays. Structured packings were installed in the wash and stripping sections. A new kerosene side draw was installed at the inlet to Tray 16, approximately mid-way between the HN and diesel draws. A side stripper was added to strip the kerosene. Initially, the kerosene rundown system was incomplete so the kerosene side draw was not used.

The naphtha end point was maintained using a temperature control (TRC) just below Tray 23. This TRC adjusted the TPA duty. The TPA was drawn from a sump just under the downcomer from Tray 21 to Tray 20, so that any liquid not drawn overflowed onto Tray 20 and became reflux to the naphtha-diesel fractionation section.

The diesel cloud point was maintained at another temperature control (TRC) just above Tray 10. This TRC adjusted the BPA duty. The BPA was drawn from a total draw chimney tray just below Tray 9. The level control on the chimney tray manipulated the wash-back which was the reflux to the packed wash bed below.

After the retray, the tower was not able to achieve the 52,500 BPSD (7,070 t/d) it had previously achieved. At throughputs exceeding 38,000 BPSD (about 5200 t/d) the tower would start to flood with a rise in pressure drop and the HN product going off-specification with heavy components. Initial gamma scans showed the tower to be flooding with the flood initiating near the kerosene draw (around Trays 14-16) and building up till Tray 23. The pumparound Trays 21-23 were jet tab trays and were not changed in the revamp.

In addition, a severe instability that we termed an "*Excursion*" would set in at about the same rates as the flooding initiated. A relatively small increase in BPA duty or reduction in coil outlet temperature (COT) would trigger a massive reduction in tray 10 temperature, usually accompanied by a HN yield loss, with the lost yield ending up in the atmospheric residue. This paper focuses on this phenomenon and its troubleshooting.

Nature of the Excursion

In a crude tower, reducing COT or increasing the BPA duty is expected to cool the tower down, leading to a distillate yield loss to the residue. This was the normal observed behavior in this crude tower. The excursion was an intensified version of this phenomenon that took place in this crude tower when operated at high loads, (above 4,800 t/d, or 36,000 BPSD), especially when the diesel make was high. There was an instantaneous large drop in Tray 10 temperature. Tray 10 temperature was measured by two temperature sensors, one connected to the controller (TRC), the other to an indicator (TI). In the absence of excursion, both TRC and TI read about 300°C. Upon the onset of excursion, the TRC dropped sharply by about 11°C, but the TI showed a much greater drop, as much as 30–40°C. The drop occurred almost instantaneously, and was usually (but not always) accompanied by loss of naphtha yield and gain in residue yield.

Historically, the TRC on Tray 10 was used by the operators to control the diesel cloud point. The TRC manipulated the heat duty of the BPA. Typically, if the operator wished to change the diesel cloud point for 0°C to −1°C, he would lower the TRC by 2°C. Following the revamp, the response of the cloud point to the TRC became erratic. A change of cloud point from −2°C to −5°C was accompanied by an excursion and a TRC drop of 11°C. Some times the TRC would jump 10 to 15°C without any change on the cloud point analyzer. The diesel cloud point could no longer be controlled from the TRC.

The large reduction in Tray 10 TRC upon excursion was usually accompanied by:

- An even larger reduction in Tray 10 TI. Typically an 11°C reduction in the TRC was accompanied by a 33°C reduction in the TI.
- A reduction in HN make and an increase in wash-back and residue make. The diesel make did not change much.
- Cooling of tower temperatures between the feed point and Tray 23.

Early Endeavors

Early simulation work suggested that at the excursion condition, there was leakage from tray 11, bypassing the diesel draw. Around 20% of the tray 11 liquid was estimated to be leaking.

In July 1993, a month after the tower was restarted with the revamped trays, it was briefly shut down again. The purpose was to seek construction anomalies that could have led both to the premature flood (discussed in a companion paper) and to the excursion. With respect to the excursion, the focus was to look for leakage around the diesel draw.

The diesel draw was water-tested and showed no leak at all. The draw was seal-welded in a previous turnaround and the seal welding was proven effective. The BPA chimney tray was also water tested and showed no leak. This tray was seal-welded during the May 1993 revamp. A few valve floats on tray 11 were found popped-out, but their fraction was far too small to explain a significant leak. The floats were reinstated and the joints of tray 11 were seal-welded at the July shutdown to minimize leakage. Some minor modifications (described below) were performed to the diesel draw.

Following the brief July shutdown the tower was returned to service with no appreciable change in performance, both on the flooding and excursion issues.

Diesel Draw Geometry and Modification

Figure 2a shows the diesel draw geometry prior to the May 1993 revamp. Tray 11 (like all the other trays) was a jet-tab tray. The tray 11 downcomer was 585 mm wide and terminated in a seal pan. The clearance under the downcomer, the seal pan weir height, and the horizontal distance between downcomer and the weir were all 50 mm. The seal pan overflowed into an 855 mm wide draw pan. The draw pan floor was 380 mm beneath the seal pan. The draw pan outlet weir was 720 mm tall, rising 290 mm above the top of the outlet weir of the seal pan. The diesel side draw was drawn by a 12-inch nozzle. The bottom of the nozzle opening was almost flush with the draw pan floor.

Figure 2b shows modifications to the downcomer from tray 11 and the seal pan during the May 1993 tray replacement revamp. The downcomer width was reduced from 585 mm to 370 mm. The downcomer clearance was increased from 50 mm to 90 mm, and the seal pan weir height was raised from 50 mm to 115 mm.

Figure 2b shows the location of the thermocouples used for the TI and the TRC on tray 10. The TI was just near the top of the draw pan overflow weir. During the brief July 1993 shutdown, there were concerns that some liquid from the draw pan or seal pan may be splashing on the TI, causing the observed large temperature dips. In an attempt to alleviate the dips, the draw pan outlet weir was raised 50 mm, from 720 to 770 mm. Further, in the region of the TI thermocouple the weir was raised by another 50 mm to 820 mm.

Theory 1: Can the Diesel Draw Experience a Hydraulic Bottleneck?

The diesel flow rate at which the excursion was experienced was about 2000 tons/day (t/d).

At this flow rate, the liquid velocity at the downcomer entrance was about 0.06 m/s, which is well within good design criteria[1] for downcomer entrance flood (or downcomer choke flood). It was concluded that downcomer choke was unlikely to cause a hydraulic bottleneck at the diesel draw downcomer.

The residence time in the downcomer was within good design practices for normal downcomers[1] , but far too small to permit adequate degassing of the downcomer liquid. This is not important for a regular downcomer, but has an impact on liquid draws. *Distillation Operation*[1] recommends a minimum residence time of 0.5-1 minute for adequate degassing of tower liquid. The modified downcomer (Figure 2b) at the most would provide about 0.2 minute of degassing time, well-short of that needed for adequate degassing. Prior to the revamp the downcomer area was twice as great, doubling the residence time to about 0.4 minute, quite close to the time needed for adequate degassing.

Without adequate degassing of the draw liquid, the diesel draw nozzle and rundown lines need to be sized for self-venting flow[1]. The correlations of Sewell[2] and Simpson[3] which have been found excellent for this type of calculation show that the 12-inch draw line could at the most handle 2000 t/d of self venting flow, which coincided with the maximum observed diesel draw rate.

This provided a simple theory that seemed to successfully explain all the observations. At diesel draw rates of less than about 2000 t/d, the diesel draw nozzle and rundown line were large enough to handle the aerated liquid flow. At higher rates, the line capacity was exceeded. Liquid would back up in the draw pan and eventually overflow onto tray 10, causing the excursion. The overflowing liquid would pour onto the TI and on the TRC, cooling them down. The TI cooled down more because it was closer to the draw pan. The excursion did not occur prior to the revamp because the downcomer from tray 11 was large enough to degas the liquid before it reached the diesel draw. While the diesel draw nozzle and diesel rundown line were inadequately sized for aerated liquid, they were large enough for (degassed) liquid flow. The excursion therefore resulted from a diesel draw bottleneck caused by the downcomer modification.

Test 1: Did the Diesel Draw Pan Overflow?

The above theory was tested by neutron-backscatter scanning the diesel draw pan at excursion conditions with the diesel draw-off rate near 2000 t/d. Figure 3 is a to-scale drawing showing the regions where the neutron scans detected liquid. The bottom 1/3 to 1/2 of the downcomer contained aerated liquid or dense froth, with lower density froth above it. There was about 300 mm of liquid or very dense froth in the diesel draw pan. The top 80 mm of this draw pan contained lower-density froth. There were 255 mm of medium density froth above the draw pan liquid in the region between the overflow weirs of the seal pan and draw pan. However, the region above this was clear so the draw pan was shown not to overflow. The draw pan overflow theory was shown to be invalid.

Theory 2: Excursion Linked to Flooding Initiating on Trays 14-16

Some people theorized that the excursion was linked to the flooding that was known to initiate on trays 14-16. Absence of flooding on trays 11-13, as verified by a multitude of gamma scans, argued against this theory, so it was considered unlikely. Nonetheless, with the leading theory of draw pan overflow hitting a dead end, there was interest in exploring this alternative theory. Good troubleshooting investigations should test both the likely and the unlikely theories(1). The proposed test had the added benefit of permitting a quantitative comparison of an initial "no excursion" steady state to a final "excursion" steady state, without being influenced by transient changes in liquid inventory on trays.

Test 2: Switching PA Duties Test

To test the flood link, the tower was set in the "no excursion" mode. Then the heat balance was changed by radically raising the BPA duty while reducing the TPA duty by an equal amount, with the throughput kept constant. This drastically lowered the vapor traffic on trays 11 – 23. The comparison between the final and initial steady states gave:

- The BPA flow was raised by 175 t/d, effecting a duty increase of 1.1 Gcal/h.
- The TPA flow was decreased, effecting a duty decrease of 1.1 Gcal/h.

- The diesel draw flow rate remained constant at 2000 t/d.
- The total HN make rate decreased by 270 t/d (down 29%).
- The wash-back flow rate increased by 330 t/d (over 25%).
- The residue flow rate increased by 250 t/d (up 13%).
- Tray 10 TRC dropped by 23°C, and the tray 10 TI by 45°C. An excursion took place.
- Temperatures in the naphtha-diesel fractionation section dropped by 11°C.
- Temperatures in the wash section dropped by 7°C.

In this test, the increase in BPA duty was matched by the reduction in TPA duty, so the net enthalpy change was the additional heat removed in the product stream. The increase in the hot residue make at the expense of the cooler HN stream represents an increase in net heat removal from the tower. This reduced the condenser duty, the top reflux, and the HN make.

This test reduced vapor loads and the proximity to flooding on trays 11-23. So if the excursion was linked to flooding, it should not have occurred in this test. The observation that it did occur disprove the link between the flood initiating on trays 14-16 and the excursion.

Theory 3: Excursion Caused by Tray 11 Leakage

The findings of Test 2, ie, that the excursion still took place when the vapor flow rate was reduced, naturally bred the hypothesis that weeping, not flooding, was the cause of excursions. A test could be devised to monitor the tray 11 leakage rate.

Since the diesel draw was a total draw, a simple mass balance on the BPA gives:

WASH-BACK = (TRAY 11 LEAK) + BPA CONDENSATION

Based on this equation, the leakage rate can be monitored by observing the wash-back flow rate while maintaining the BPA condensation rate (ie, the BPA heat duty) constant.

Test 3: Tray 11 Leakage Test

This test started with the tower at the excursion mode. The coil outlet temperature was raised from 365°C to 370°C. The BPA flow was manipulated to keep BPA duty constant.

As the COT rose, the wash-back dropped by 160 t/d, the residue make dropped by 100 t/d. The HN make rose by 120 t/d. Tray 10 TRC rose 14°C, and tray 10 TI rose 35°C. Temperatures on trays 16 and 19 rose about 5°C.

The above equation shows that, in the absence of tray 11 leakage, keeping the BPA duty constant would have assured constant wash-back rate. The large change in wash-back flow rate (which took place upon excursion) at no change in BPA duty demonstrated that the excursion was due to change in leakage from tray 11.

Theory 4: Excursion Caused by Tray 11 Weep

The verification that the excursion was due to leakage from tray 11 gave rise to a new theory. It postulated that liquid weep from tray 11 rained on the TI and TRC, causing these temperatures to drop. As COT went up, the weep stopped. The TI and TRC dried, causing these temperature measurements to rise. The different cooling rates of the TI and TRC can be explained by different amounts of weep at each location, more at the TI where the vapor rate was lower due to the draw pan obstructing the vapor flow. Also, as the weep stopped, there was less wash-back and less resid make.

This theory explained all the observations. Its only weak link was its difficulty of explaining the suddenness of the change. Also, like any theory, it needed testing.

Test 4: Tray 11 Weep Theory Testing

Work by Lockett and Banik(4) and Colwell and O'Bara(5) showed that tray weep rate is a primary function of the vapor hole velocity. It also showed that the weep rate is a much weaker function of liquid flow rate. It therefore makes sense that small changes in COT, which change the vapor hole velocity, can get tray 11 in and out of weeping (and therefore, excursion) relatively fast. In contrast, based on the above work, the weep rate should be rather insensitive to changes in liquid load. A "devil's advocate" test would therefore monitor the response of the excursion to changes in liquid load at constant vapor rate. A minor or no response of the excursion to mild changes in the liquid load would validate the tray 11 weep theory above.

The tower was set in the excursion mode. Vapor rate to tray 11 was kept constant keeping both the BPA duty and the COT constant. The TPA duty was then reduced by switching off a fan on the TPA air cooler. This change only reduced the tray 11 liquid flow rate, and was expected to have no impact on the excursion.

The result was astonishing. The liquid load reduction took the tower out of excursion. The TRC and TI on tray 10 rose 12°C and 35°C respectively. The wash-back rate dropped 260 t/d. The residue make dropped 100 t/d, and the HN make rose 140 t/d. The diesel make dropped 50 t/d and the Light Virgin Naphtha (LVN) make from the tower overhead dropped 30 t/d.

This result disproved the tray 11 weep theory. This test also established that the excursion appeared very sensitive to the HN draw rate. This implied a liquid-load-sensitive bottleneck, not a vapor-load-sensitive one.

This test sent us back to reconsider a diesel draw bottleneck as we initially considered (Theory 1), and discount the possibility of a tray weep bottleneck. Further testing was needed to verify the liquid dependence.

Test 5: Kerosene Mode Tests

A very interesting operating experience in the crude unit was that excursions were not experienced while the tower operated in the "kerosene mode"—i.e., when the kerosene side-draw was operated. This combined well with the conclusion that the excursion was associated with a diesel draw limit. When kerosene was drawn, the diesel make was lower, and the draw bottleneck would not be encountered.

To further test the now-revived diesel draw bottleneck theory, we attempted to force an excursion whilst in the kerosene mode by lowering the COT. When in the no-kerosene mode, as COT was dropped below 363°C the tower seldom got out of excursion.

Test 5A

The tower was set in the kerosene mode, with 350 t/d of kerosene drawn. Initially, the tower operated in a non-excursion mode with the COT at 371°C. The COT was slowly lowered to 357°C. The BPA duty was kept constant, and so was the tower reflux. The TPA duty was allowed to decrease to maintain the heat balance.

No excursion occurred. Tray 10 TRC and TI both went down by 15°C. The wash-back remained constant. The diesel make dropped 175 t/d, with the residue make rising 250 t/d. The TPA duty went down by 2 Gcal/h.

This test demonstrated that there was very little weep from tray 11, and confirmed that unloading the diesel draw kept the tower out of excursion.

Test 5B

While doing the above tests, at different times, we shut (or nearly shut) the kerosene draw-off valve. We could only do this for 0.5-1 minute because the stripper had only a small surge volume. Table 1 shows the results. Excursion was immediate upon valve closure, but only when the diesel plus kerosene flows exceeded 1960 t/d.

This result demonstrated that the bottleneck occurs at a sharp diesel draw rate of 1955 – 1985 t/d. It conclusively showed that the diesel draw rate was the only motive for going in and out of excursion. With liquid having a typical time constant of 0.1 minute per tray, and there were only a few trays between the kerosene and diesel draws, the rapid response was explained.

The kerosene tests proved that the excursion was due to a hydraulic limit at the diesel draw at about 1970 t/d.

Table 1. Tests of Shutting the Kero Drawoff

COT, °C	Diesel Draw, t/d	Diesel Draw when Kero Draw Stopped, t/d	Excursion?
371	1650	2000	Yes, twice
368	1620	1985	Yes, little
363	1600	1955	No sign
357	1490	1840	No sign

Excursion Cause

Our extensive field tests took us back to the very beginning, when we suspected a hydraulic limit at the diesel draw at about 2000 t/d. At this draw rate there was a switch from self-venting flow to choked flow at the diesel draw nozzle and rundown line. The tests provided conclusive evidence supporting the existence of this bottleneck.

The limit was not encountered prior to the revamp because of the downcomer from Tray 11 provided twice the degassing time. This larger (0.4 minute) degassing time was probably sufficient to eliminate the aeration from the diesel draw, so the diesel draw nozzle and rundown line did not need to handle self-venting flow. Both were sufficiently large to handle the *non-aerated* diesel draw liquid flow rate.

At the revamp, the downcomer volume was approximately halved and so was the degassing time. At 0.2 minute residence time, there is little doubt that the draw liquid was not fully degassed, and some aeration remained. The diesel draw nozzle and rundown line could handle up to about 2000 t/d of aerated liquid draw without encountering a limit. Once pushed to higher rates, the gas bubbles would choke the line, causing liquid backup in the diesel draw pan.

The large drop in Tray 10 TI and TRC upon excursion was due to rain from the diesel draw pan. Since the TI was closer to the draw pan than the TRC (Figure 2b), it got wetter and showed a much greater temperature drop. Below the limiting diesel draw rate (about 2000 t/d), there was no rain, and both the TI and TRC read the same higher temperature.

The above elements were included in our initial theory (see earlier under "*Theory1: Can the Diesel Draw Experience a Hydraulic Bottleneck*"). The one other element included in the earlier theory is that the rain was generated by liquid backup and overflow of the diesel draw pan. This element was disproved by the neutron scans that showed no overflow (Test 1).

It is important to recognize that neutron backscatter scans do not penetrate deep into the vessel. They provide excellent measurement of liquid levels and froth densities near the tower wall, but only penetrate about 200 mm into the vessel. So Figure 3, based on the neutron scans, only shows the liquid and froth profile close to the tower wall.

There were several gamma scans shot in the tower. One scan chord ("B") passed parallel to the diesel draw pan wall and about 60 mm away from the pan wall (Figure 4). This chord detected a significant quantity of liquid in the vapor space right above the draw pan wall, even under non-excursion conditions. The quantity of liquid seen at the same scan chord was much higher in the excursion scans (Figure 4). Other active area chords shot in the tower did not detect liquid at the same elevation (just below tray 11) even at excursion conditions. The gamma scans therefore show that at the excursion condition there was either extensive entrainment from the diesel draw pan, or extensive weep from tray 11, right above the draw pan wall. The weep could have been induced by the low vapor upward velocity in that region due to the obstruction by the draw pan. Tray 11 was equipped with the low pressure drop venturi valves, which are less leak-resistant than conventional sharp-orifice valves, and conducive to weeping.

The last remaining question is why it rained on the TI and TRC during the excursion and not under non-excursion conditions. As has been established above, under non-excursion conditions, the diesel draw operated under self-venting flow. The aerated liquid comfortably descended, leaving a 75 mm vapor space between the top of the diesel draw liquid level and the seal pan above (Figure 5a). Vapor returning from the rundown line could spread horizontally through this vapor space, generating low vapor velocities. Neither the weep, nor the draw pan liquid, were entrained by the low-velocity vapor.

Once the self-venting flow limit would be reached in the diesel draw nozzle and rundown line, liquid/froth would back up in the diesel draw pan, eliminating the vapor space between the liquid in the draw pan and the floor of the seal pan above (Figure 5b). From now on the vapor returning from the rundown line would return as a high-velocity jet right near the nozzle. This jet either lifted liquid from the draw pan, or picked up the weep from tray 11, shooting them at the tray 10 TI and TRC. This caused the excursion. The success of subsequent modifications (and implications of Test 4) suggest that tray 11 weep may not have been a significant contributor.

Solution

The same field tests that diagnosed the cause of excursion taught us how to circumvent excursions: by operating in the kerosene mode. Following the tests, the kerosene mode became the normal mode of operation. There was little economic penalty to drawing a relatively small amount of kerosene product stream, as it could have been blended with the diesel when not desired as a product.

Operating at the kerosene mode circumvented not only the diesel draw bottleneck, but also the flooding bottleneck (which will be discussed in a companion paper). So by drawing a small kerosene product stream, both bottlenecks were overcome, and the tower feed rates could be raised from 5,200 t/d to 6,500 t/d, much closer to the design 7,070 t/d. At 6,500 t/d, the flood limit re-occurred but the excursion did not. At 6,500 t/d charge rate, the capacity loss was tolerable and an imminent shutdown was avoided.

For a permanent fix, the diesel draw from tray 11 was replaced by a seal-welded chimney tray. The diesel draw and diesel rundown line were increased to 16 inches diameter. Tray changes were also implemented to address the flood bottleneck and these will be discussed in a companion paper.

Following these hardware modifications, the tower continuously operated well at a feed capacity of 7,070 t/d (52,500 BPSD), with or without the kerosene side draw. The tower was tested and shown to operate well at a feed capacity as high as 7,750 t/d (57,500 BPSD). The high capacity trays (which were the same type, but different design, compared to the originals) well-exceeded their original design capacity. Excursions no longer occurred, soon to become a faded memory.

References

1. Kister, H. Z., "Distillation Operation", McGraw-Hill, NY, 1990.

2. Sewell, A., "Practical Aspects of Distillation Column Design", The Chem. Engineer, No. 299/300, p. 442, 1975.

3. Simpson, L. L., "Sizing Piping for Process Plants", Chem. Eng., p.192, June 17, 1968.

4. Lockett, M. J., and S. Banik, Ind. Eng. Chem. Proc. Des. Dev.,25, p.561, 1986.

5. Colwell, C. J., and J. T. O'Bara, Paper presented at the AIChE Spring National Meeting, Houston, Texas, April 1989.

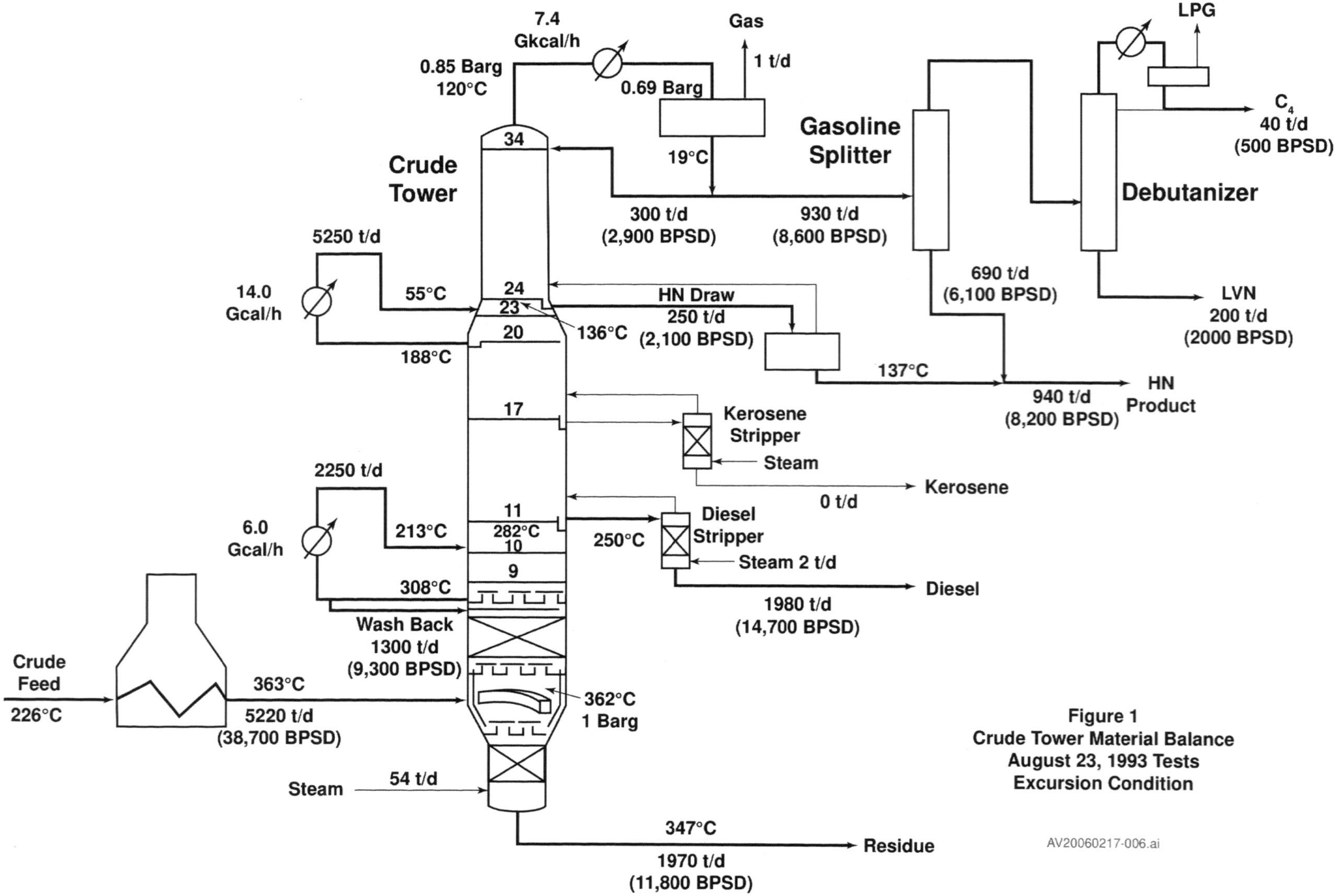

Figure 1
Crude Tower Material Balance
August 23, 1993 Tests
Excursion Condition

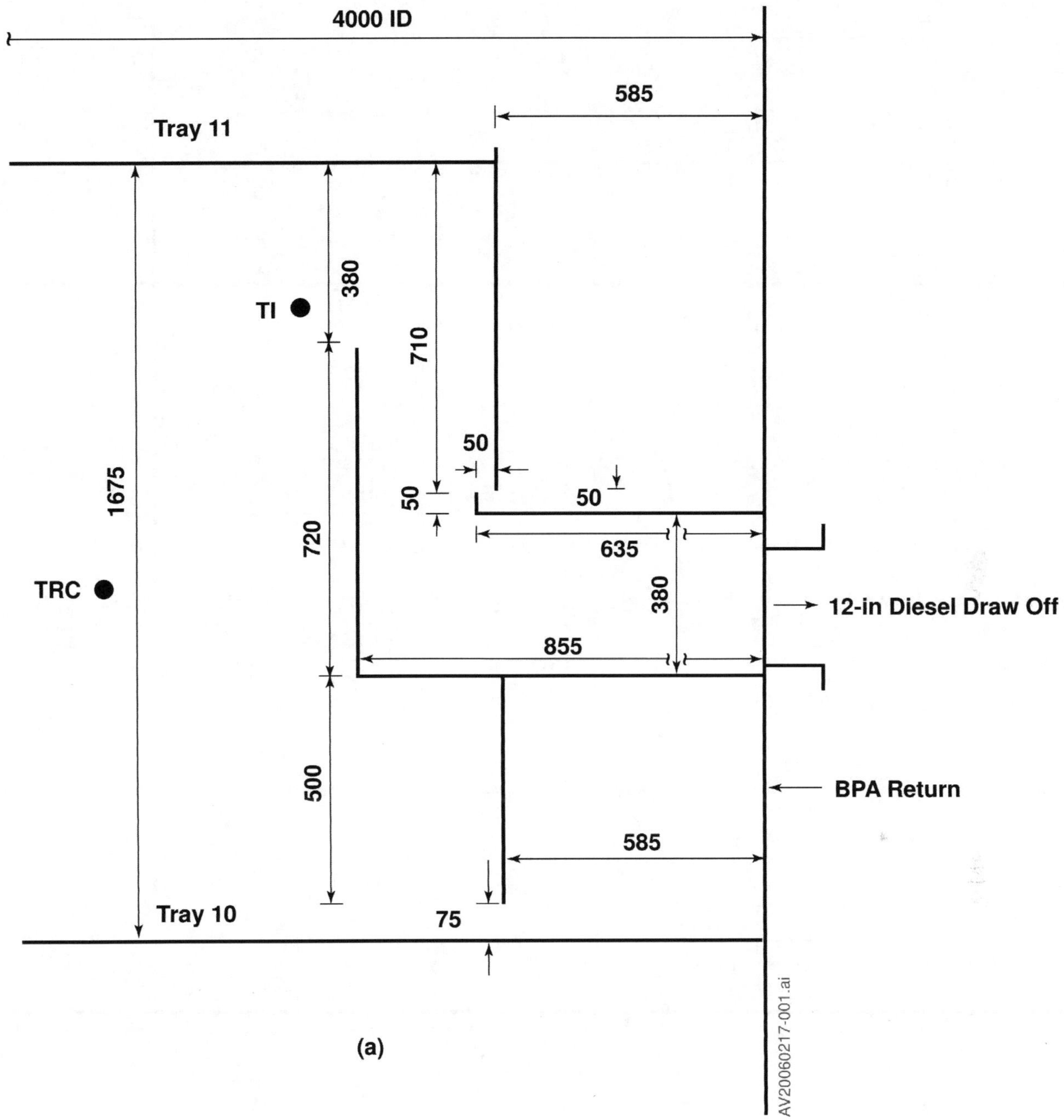

Figure 2 - Diesel Draw Off
(a) Prior to Revamp
(b) Revamp Modification (shown in dashed lines).
Locations of TI and TRC Points are approximate.
All dimensions are in mm.

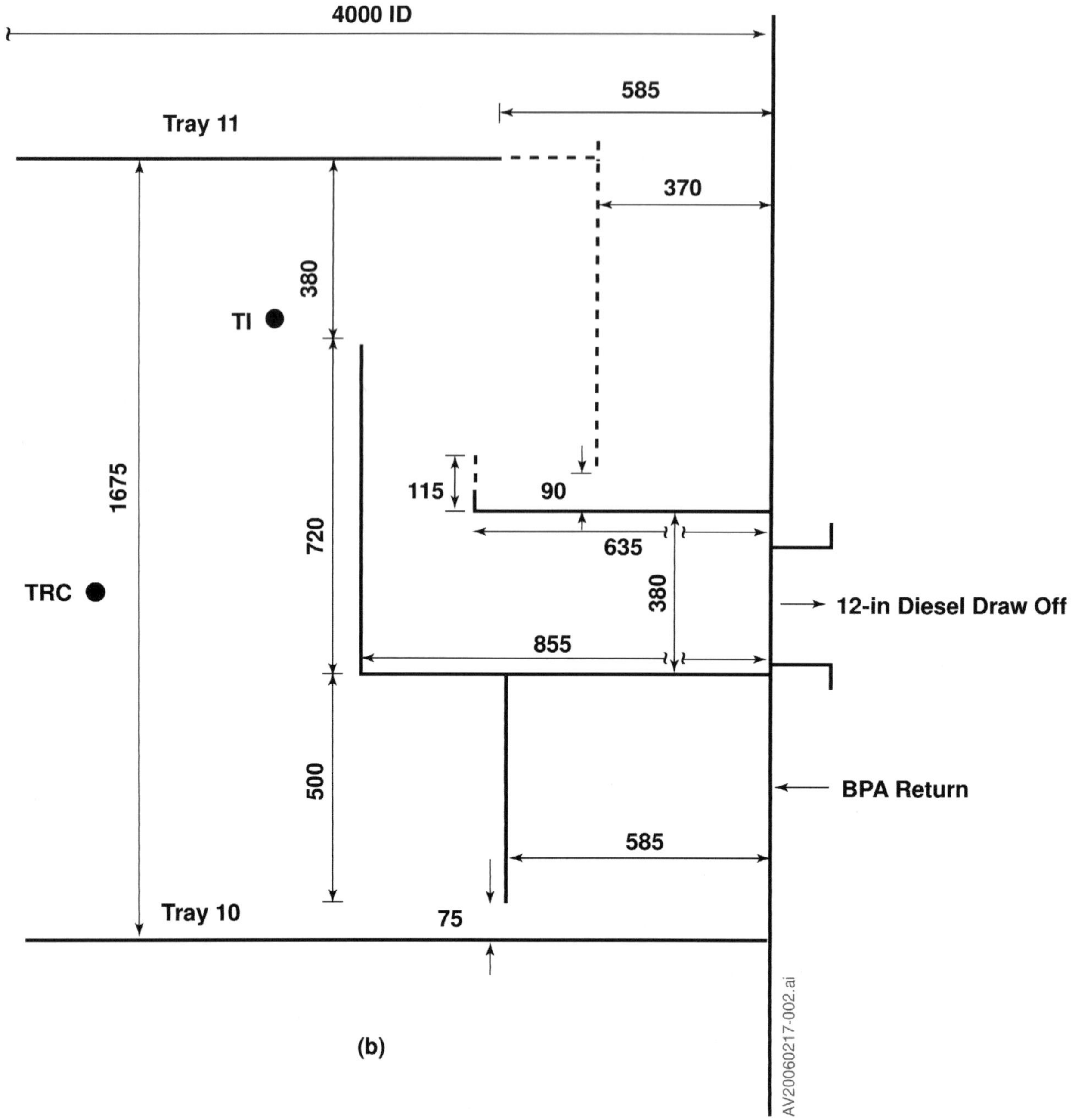
4000 ID
585
Tray 11
370
380
TI
1675
115
90
720
635
TRC
380
12-in Diesel Draw Off
855
500
BPA Return
585
Tray 10
75
AV20060217-002.ai

(b)

Figure 3 - Regions of Liquid as Seen by the Draw Pan Newton Backscatter Scans

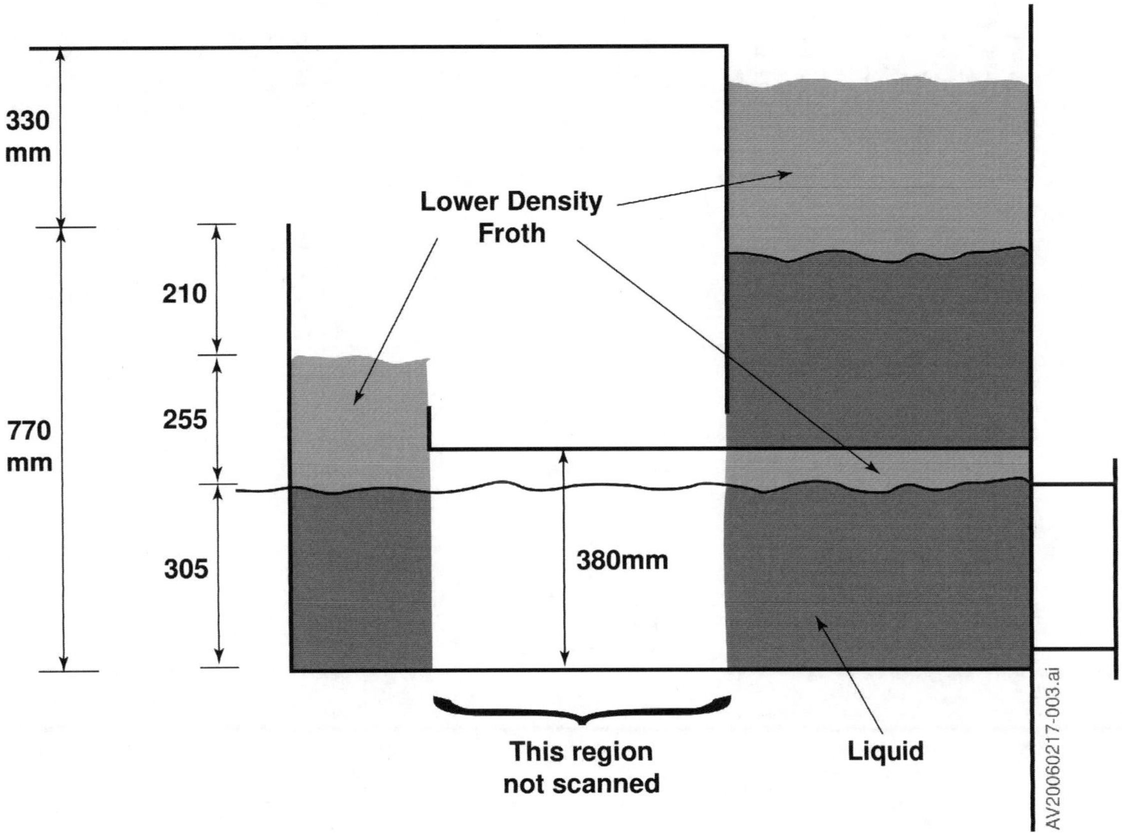

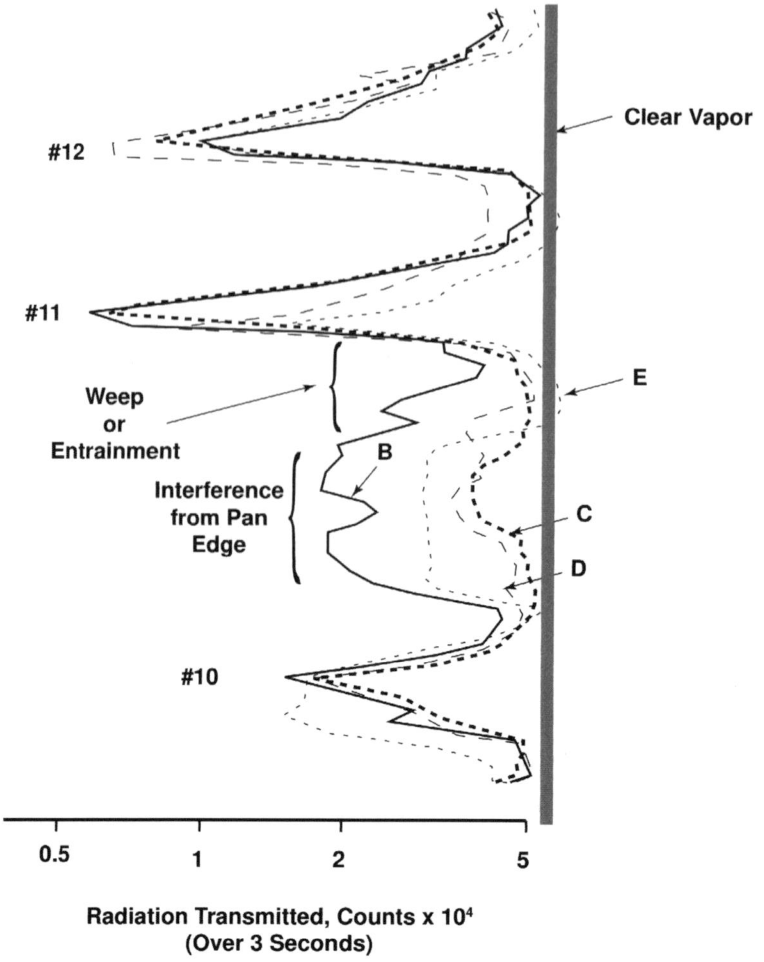

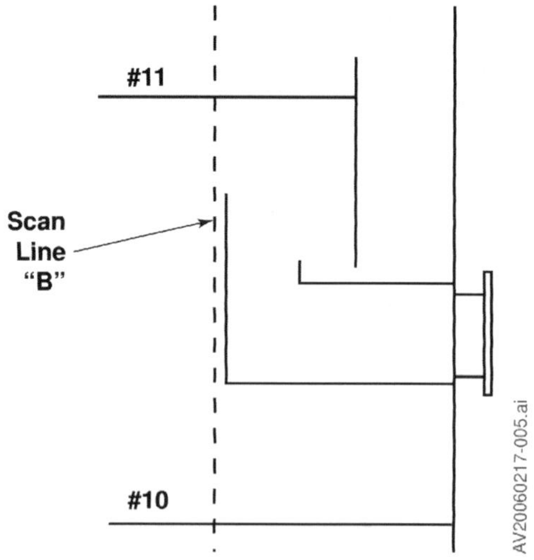

Figure 4 - Gamma scan during excursion, showing presence of liquid above the diesel draw pan. Only chord B, right next to draw pan shows this liquid. Chords C, D, E, other active area scans, show no liquid in the vapor space.

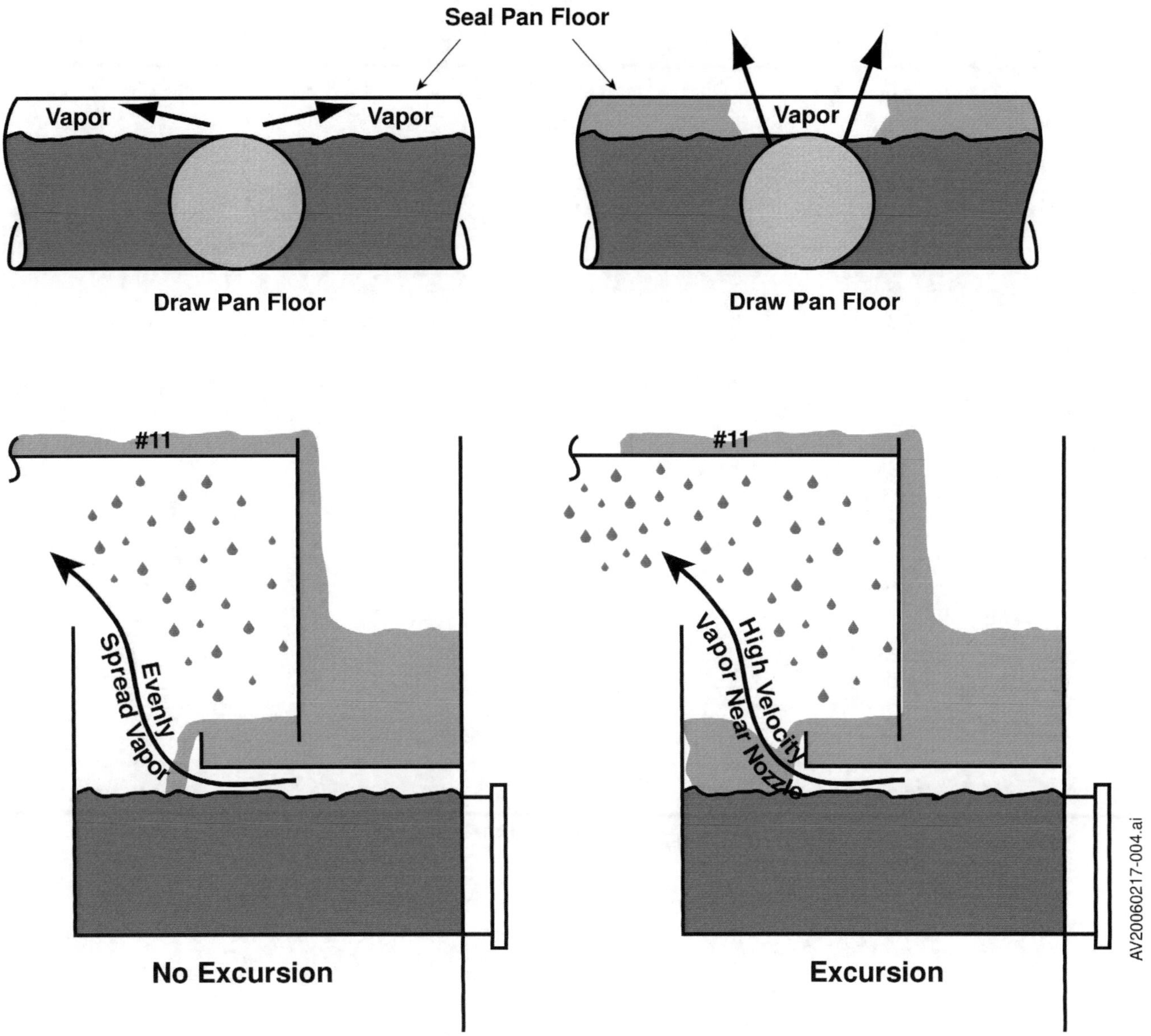

Figure 5 - Excursion Theory

PREFLASH DRUM WHEN PROCESSING HEAVY OILS: PARADOX OR REALITY ?

Silvia Waintraub[1]
Raquel Campos Cauby Coutinho[1]
Rogério Oliveira Espósito[1]

[1]PETROBRAS/RESEARCH CENTER-CENPES

Rio de Janeiro
Brazil

Presented in the Distillation Topical Conference of the 2007 Spring AIChE Meeting
Houston, April 22 - 26

March/2007

Unpublished

Abstract

The first thinking when processing heavy oils is not to use either a preflash drum or a preflash tower, as the amount of light hydrocarbons (combustible gas, LPG and light naphtha) tends to be very low. With this configuration it is necessary to prevent vaporization along the preheating train and before the atmospheric furnace pass inlet valves. On the other hand, two aspects appear related to heavy oil processing. The first one is that desalting is more difficult, requiring higher temperatures. Desalted crude water content is higher due to the effect of higher water solubility, higher content of surfactants, and higher conductivity among other effects that lower desalting efficiency. The second one is that very high inlet atmospheric furnace temperature can be achieved due to greater heavy products yields and the use of pinch technology to reduce energy consumption. The result of the combination of high temperature and higher salty water content in the desalted crude is a necessity of a very high operating pressure to prevent vaporization and sometimes a special metallurgy. In some cases it would be required even to change equipment and piping to a more expensive pressure class. Thus, as a consequence, one solution is to use a configuration with a preflash drum to allow the vaporization of the water and some amount of light hydrocarbons, reducing the vapor pressure at the furnace inlet, and eliminating any aqueous phase at the preheating trains after desalters. In this paper, a case study of a grass-roots refinery design having a 16 °API gravity feedstock will be presented. Some different oil and water mixtures had those vapor pressures studied using process simulators and compared with PVT cell experimental results.

Introduction

In general, distillation units may have two quite different configurations, depending on the presence or not of the preflash tower (or, in some cases, a preflash drum).

In the first case, the crude preheating train is designed in order to produce the maximum possible vaporization at the entrance of the preflash tower. This kind of configuration is presented in Figure 1.

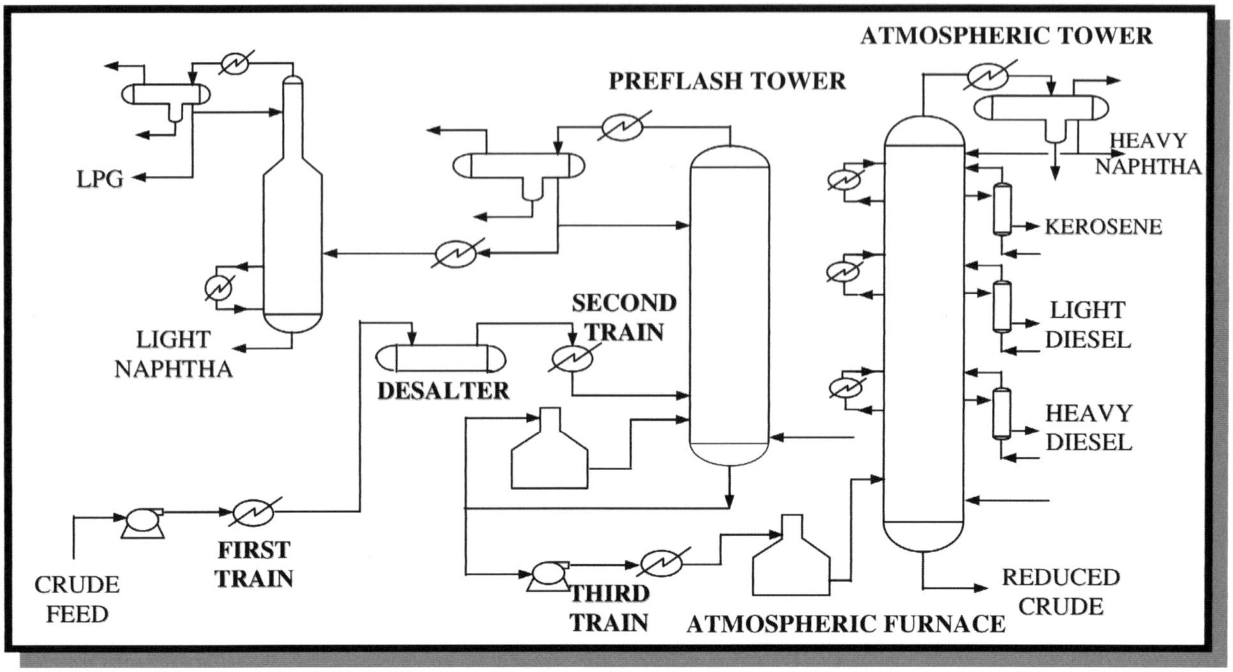

Figure 1: Configuration with Preflash Tower

On the other hand, in the configuration without the preflash tower, as showed in Figure 2, the outlet pressure at the last heat exchanger of the crude preheating train should be high enough to keep the petroleum in liquid phase, allowing a good controllability at the pass inlet valves of the atmospheric furnace.

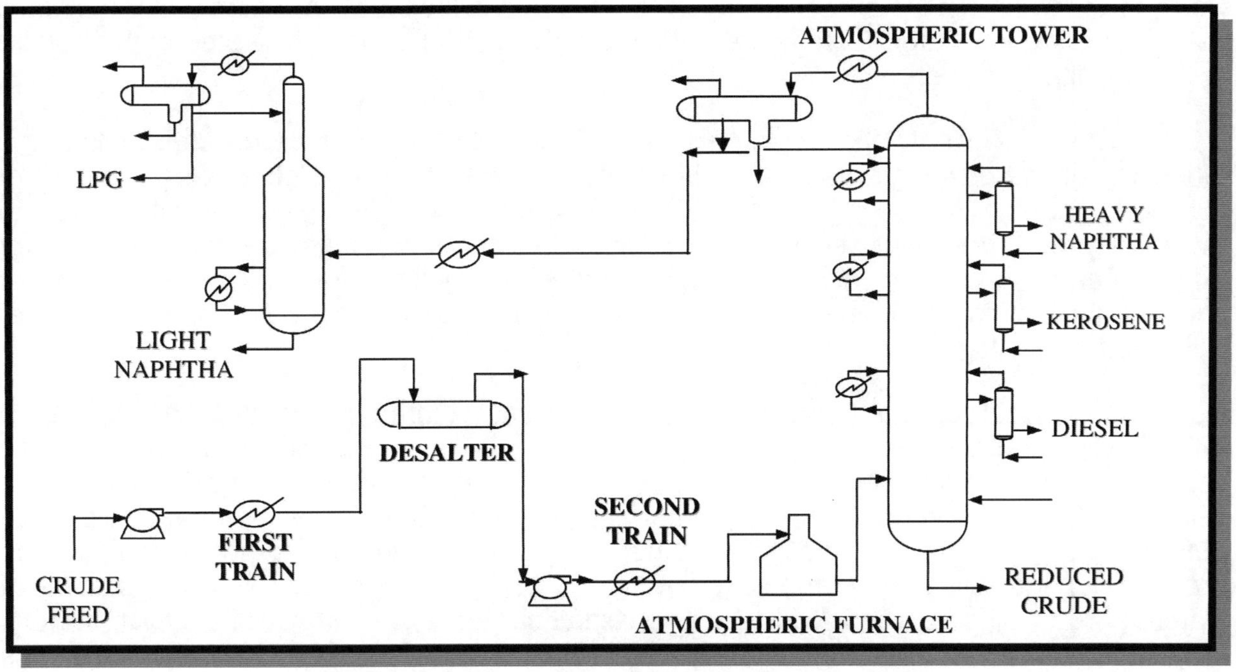

Figure 2: Configuration without Preflash Tower

Previous studies (Sloley, 2001) demonstrated the influence of using either a preflash drum or a preflash tower in shifting material upstream, and the change in material and energy balance of the atmospheric tower, having an influence in the diesel cut point. He emphasized that normally these equipment are used in revamps because in a grass-roots plant it would have few, if any, advantages, and often would cost more than a plant with the same capabilities and without a preflash drum or tower.

Golden (2005) mentioned that preflash drums or towers are a debottlenecking tool to increase crude unit capacity, with high capital cost and prone to foaming.

So, the first choice of configuration for a new design when processing heavy crude seems to be the second one mentioned above (without preflash), as the amount of light hydrocarbons (combustible gas, LPG and light naphtha) in such oil is very low.

In fact, some others aspects should be kept in mind when choosing a distillation configuration. This paper discuss some reasons why in a grass-roots refinery, processing heavy and high naphthenic acid crude oil (16°API and 2.3 mgKOH/g), adding a preflash drum is attractive and cost saving, in an opposite decision of the above discussion.

Desalting Heavy Crude Oil

The purpose of the desalting process is to remove salt, water and fine solids suspended in the crude oil. The principle of the desalting operation is to wash the crude oil, under controlled conditions, to dissolve salts, dilute residual water from production fields and to wet impurities. Usually, electrocoalescence is considered to be the best method to enhance the water and oil separation in the refinery environment. With this technique the water in crude oil dispersion/emulsion is resolved by coalescence in a high voltage electric field. The enlarged water droplets are settled by gravity and brine is collected in the bottom of the desalter vessels.

One of the most challenging problems faced when refining heavy crude oils is the efficiency of the desalting process. This is one of the factors that influence the extent of corrosion damage to process equipment, to crude unit run lengths and also contributes to the quality of the end products. Furthermore, the increasingly severe requirements for environmental protection and the urgent need of minimization of the fresh water consumption reinforce the major attention that has been given to the improvement of pre-treatment technology.

The water-oil emulsion stability is a critical element affecting the operating efficiency of heavy crude oil pre-treatment. Crude oils have natural emulsifiers, such as asphaltenes and resins (including naphthenic acids) that contribute to the formation of a viscoelastic film on the crude oil-water interface (Aske et al., 2002, McLean and Kilpatrick, 1997a,b). Besides, resins and asphaltenes aggregates interaction, additionally to the presence of fine solids (both organic and inorganic), play an important role in the stability of water-in-crude oil emulsions (Sullivan and Kilpatrick, 2002). In heavy oils, which contain large amounts of surfactants, this stabilization mechanism is more significant and brings severe consequences.

Desalting pilot plant studies and industry scale observations show that process efficiency depends largely on viscosity and density, but some other aspects like surfactants and solids content also affect (Coutinho, 2005).

Heavy oils have:

- High viscosities;
- High specific gravity, and as a consequence lower differential density related to water;
- High amount of emulsifying components, such as asphaltenes, resins, solids and naphthenics acids;
- A tendency of solids build-up at the water-oil interface.

Figure 3 shows the differential density between crude oils and salty water solutions.

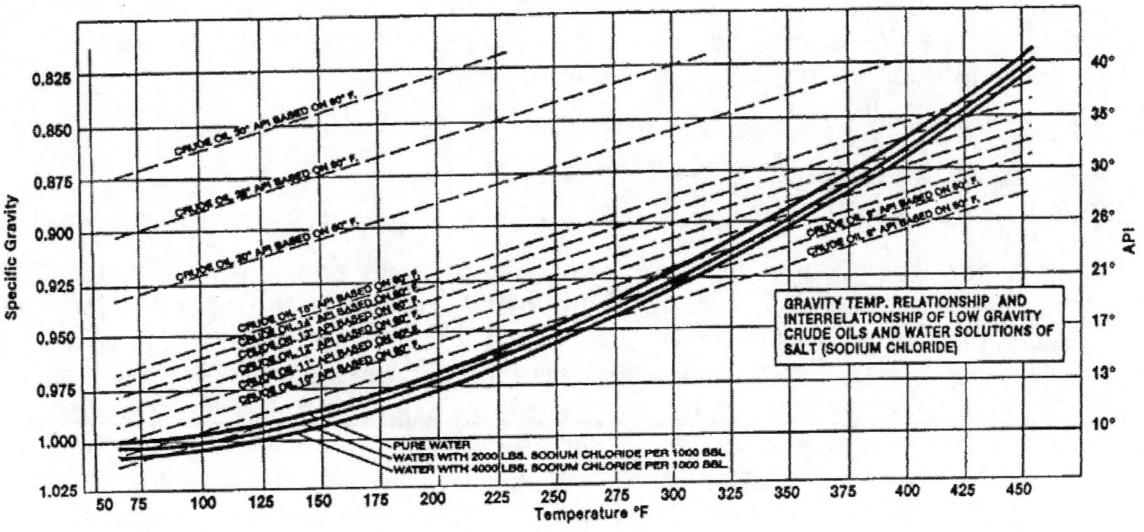

Figure 3: Differential Density of Low Gravity Crude Oils and Salty Water Solutions (Warren, 2002)

As a consequence of these characteristics, one of the major problems with heavy oil desalting is the formation of a very stable water-oil emulsion. The presence of naphthenic acids and their salts tend to increase this stability (Coutinho, 2005).

As desalting is more difficult, higher temperatures are required. On the other hand, with the increase in temperature, there is an increase in water solubility, as showed in Figure 4, and also a reduction of desalting efficiency due to the increase of electrical conductivity of crude oil causing a decrease of the electric field between the electrodes. The temperature limits are also set by the insulators and bushings resins.

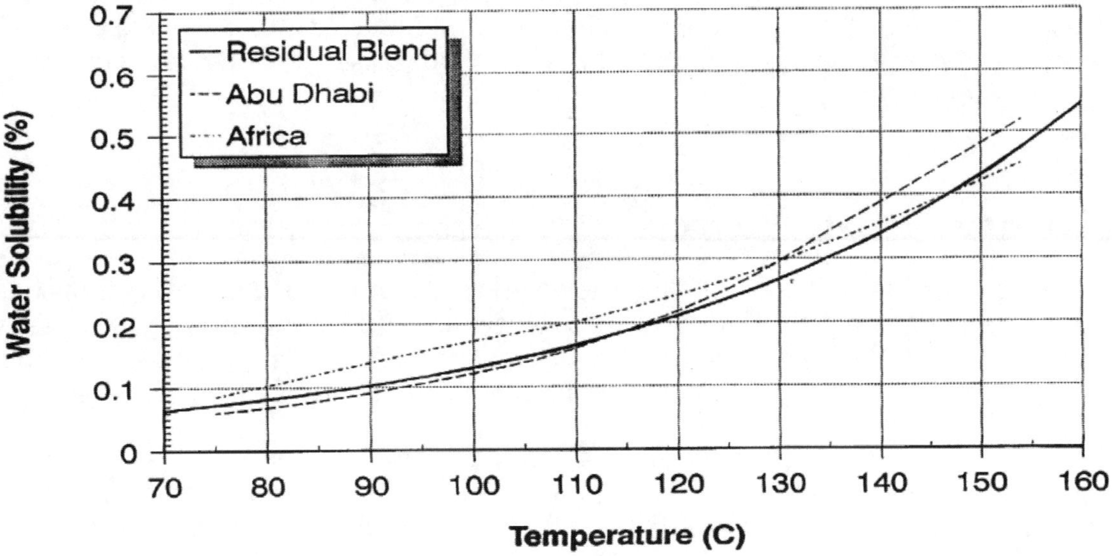

Figure 4: Temperature Influence in Water Solubility in Oils (Warren, 2002)

The consequence of all discussed in this item, is that desalted crude water content tends to be greater for a heavy crude oil than for light ones and common BS&W (Basic Sediments and Water) values will be in a range of 0.4 to 0.8.

Preheating Train Design

In a new design, the number of heat exchangers and stream matches can be determined by using Pinch Technology methodology. This approach searches for reducing energy consumption, through minimization of cold and hot utilities, with an efficient heat recovery between streams, for an attractive investment in the network. The first step, called target, is the determination of minimum temperature approach for the train synthesis, where annual equipment and energy costs are minimized. In the second step, called network synthesis, the matches between cold and hot process streams are defined. The number of pumparounds and their duties can be chosen by integrating the pinch analysis with atmospheric column simulation, using restrictions of L/V minimum values for a given product specification.

Another way to optimize heat recovery is the use of Compact Plate Heat Exchangers, mainly when special materials are needed to handle naphthenic corrosion, where their prices become attractive. This type of exchanger is being used in many applications in refinery preheating trains (Andersson, 2007; Spangler and alii, 2006) and has some advantages when compared with traditional shell and tubes, like smaller temperature approaches and size.

As a result of high energy integration due to pinch technology and the use of efficient heat exchangers type, there is a reduction in heater fuel consumption and emissions. Crude is heated to very high temperatures at the end of preheating train (higher than 300 °C), increasing the inlet atmospheric furnace temperature. On the other hand, to keep atmospheric furnace efficiency, it is necessary to have heat recovery either at the convection section (such as steam generation and/or steam superheating) and/or installing air preheating system. An integrated study has to be done, in order to determine the minimum fuel consumption.

Determining the Operation Pressure

When there is neither a preflash tower nor a drum, the preheating train has to operate without vaporization. Pressure has to be chosen in order to guarantee that petroleum will be in liquid phase at the preheating train outlet temperature. This is necessary in order to have a good control at the atmospheric furnace inlet pass valves.

Waintraub and alii (2001) showed a real case study, where the greatest problem was the control of the atmospheric furnace, which became unstable because of vaporization before the inlet control valves. The main problem was the impossibility of controlling the rate splitting of a two phase flow at the inlet of the atmospheric furnace.

Bubble point pressures calculated by simulation depend on the modeling of the crude light ends and the amount of water. It is necessary to do the correction of the light ends in order to match the beginning of the crude TBP distillation curve with the results of the chromatography presented in the crude assay.

In Table 1a and 1b the bubble point pressures simulation results for a Light Brazilian Crude Oil (28 °API) and a Heavy Brazilian Crude Oil (16 °API) without water are presented. In these tables the results are using Graysson-Streed Equation of State (EOS) and considering light ends match or not. It can be noticed that with light ends match, the bubble point pressures are much higher.

Table 1: Bubble point pressures (kgf/cm²a) obtained by process simulator (a) Light Brazilian Crude Oil (b) Heavy Brazilian Crude Oil

	Light-ends match	
T (°C)	yes	no
50	1.2	0.42
80	1.9	0.90
100	2.5	1.4
125	3.4	2.3
150	4.5	3.4

(a)

	Light-ends match	
T (°C)	yes	no
50	1.0	0.14
80	1.3	0.29
100	1.6	0.46
125	1.9	0.75
150	2.3	1.2

(b)

In Table 2 and in Table 3 the simulation results of bubble point pressures for different water content, for the Light Brazilian Crude Oil, are tabulated. In Table 2 the results are using Grayson-Streed EOS, while in Table 3 Soave-Redlich-Kwong was used.

Table 2: Influence of water content on bubble point pressures obtained by process simulator for light Brazilian Crude Oil with Grayson-Streed EOS

	Bubble point pressures (kgf/cm²a)						
T (°C)	Water content (%v/v)						
	0	0.2	0.3	0.5	0.6	0.8	1.0
50	1.2	1.3	1.3	1.3	1.3	1.3	1.3
80	1.9	2.4	2.4	2.4	2.4	2.4	2.4
100	2.5	3.5	3.5	3.5	3.5	3.5	3.5
125	3.4	5.8	5.8	5.8	5.8	5.8	5.8
150	4.5	8.5	9.5	9.5	9.5	9.5	9.5
200	7.5	11.2	13.1	16.7	18.4	21.9	24.5
250	11.6	15.6	17.5	21.3	23.1	26.8	30.3
300	16.9	22.8	25.7	31.5	34.3	39.9	45.5

Table 3: Influence of water content on bubble point pressures obtained by process simulator for Light Brazilian Crude Oil with Soave-Redlich-Kwong EOS

T (°C)	Bubble point pressures (kgf/cm²a)						
	Water content (%v/v)						
	0	0.2	0.3	0.5	0.6	0.8	1.0
50	1.1	1.2	1.2	1.2	1.2	1.2	1.2
80	1.8	2.2	2.2	2.2	2.2	2.2	2.2
100	2.4	3.4	3.4	3.4	3.4	3.4	3.4
125	3.3	5.6	5.6	5.6	5.6	5.6	5.6
150	4.5	8.1	9.3	9.3	9.3	9.3	9.3
200	7.6	12.1	14.3	18.6	20.6	23.8	23.8
250	11.9	17.0	19.5	24.3	26.7	31.3	35.8
300	17.3	22.7	25.3	30.4	32.9	37.9	42.6

In Table 4 simulation results of bubble point pressures for different water content, for the Heavy Brazilian Crude Oil using Grayson-Streed EOS are tabulated.

Table 4: Influence of water content on bubble point pressures obtained by process simulator for Heavy Brazilian Crude Oil with Grayson-Streed EOS

T (°C)	Bubble point pressures (kgf/cm²a)						
	Water content (%v/v)						
	0	0.2	0.3	0.5	0.6	0.8	1.0
50	1.0	1.1	1.1	1.1	1.1	1.1	1.1
80	1.3	1.8	1.8	1.8	1.8	1.8	1.8
100	1.6	2.6	2.6	2.6	2.6	2.6	2.6
125	1.9	4.3	4.3	4.3	4.3	4.3	4.3
150	2.3	6.5	7.2	7.2	7.2	7.2	7.2
200	3.4	7.5	9.5	13.3	15.2	18.8	19.4
250	4.9	9.2	11.2	15.2	17.1	20.9	24.4
300	7.0	13.2	16.1	21.9	24.7	30.2	35.5

It can be observed from Tables 2, 3 and 4 an increase in bubble point pressures with the increase in water content, for a fixed temperature, up to the point where water in crude oil solubility is achieved. After this point, bubble point pressure is constant, and all the increase in water amount goes to the second liquid phase. In addition, it can be seen that water influence on bubble point pressures is much larger for higher temperatures, where the solubility is also higher. It is possible to conclude, as well, the critical influence of desalting/dehydration efficiency over the furnace operational stability.

As discussed in the previous items, high inlet atmospheric furnace temperature and high values of water content in the desalted crude tends to be a reality when processing heavy oils. The result of the combination of these two aspects is a necessity of a very high operating pressure to prevent vaporization. In some cases it would be required even to change equipment and piping to a more expensive pressure class.

Another point is that processing high naphthenic acid crude oils in these circumstances, that means, high temperature and remained chlorides in an aqueous phase, requires a very special and expensive metallurgy.

Obtaining Experimental Vapor Pressures

In order to choose the right operating pressure of the preheating train and confirm some simulation results, experimental values of bubble point pressures were determined.

The experimental procedure consists in filling a metallic pressure vessel, called PVT (Pressure-Volume-Temperature) cell, with the sample and stabilizing the temperature in a high pressure, where there is guarantee that there is only liquid phase. Pressure is then reduced until the first bubble is noticed, as a visual observation. This value is called the bubble point pressure. The operating temperature limit was of 150°C. The apparatus is showed in Figure 5.

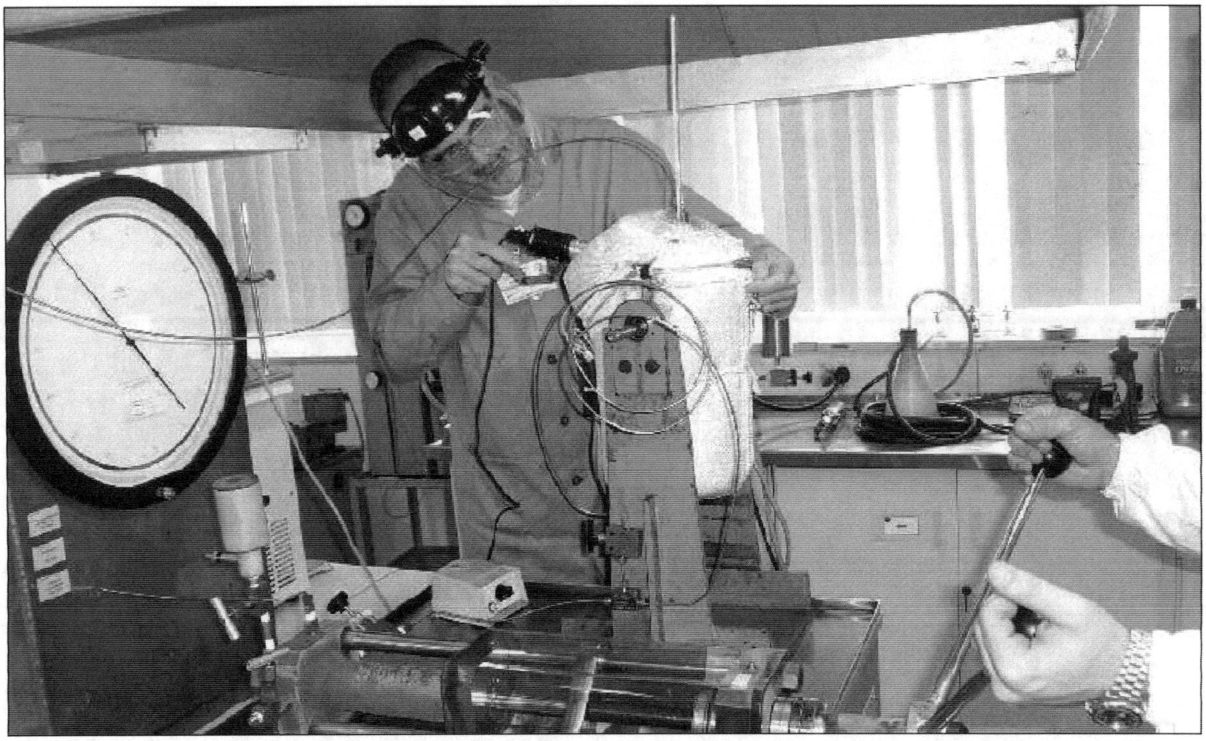

Figure 5: Determination of experimental bubble point pressures in PVT cell

Samples of the Light Brazilian Crude Oil were used. Two sets of experiments were done. The first set with dehydrated crude oil and the second with 0.6%v/v of water content. The experimental results for bubble point pressures and simulation results with Grayson-Streed EOS are shown in Table 5.

Table 5: Bubble point pressures obtained in PVT cell and with simulation for the Light Brazilian Crude Oil

T(°C)	Bubble point pressures (kgf/cm²a)			
	Water content (%v/v)			
	0		0.6	
	Experimental	Simulation	Experimental	Simulation
50	1.5	1.2	1.5	1.3
80	2.0	1.9	2.0	2.4
100	2.5	2.5	3.0	3.5
125	3.5	3.4	5.0	5.8
150	4.5	4.5	7.0	9.5
300	-	16.9	-	34.4
310	-	19.1	-	38.4
320	-	19.4	-	42.9

The predictions for the bubble pressure point for the dehydrated sample were very good, while for the sample with the presence of water was poor, but conservative.

Including a Preflash Drum

For the present case study, that the basis is the processing of heavy and high naphthenic acid crude oil (16°API and 2.3 mgKOH/g), considering a preheating train achieving 310°C at the furnace inlet and a typical water content of 0.6%v/v at desalted crude, the design pressure for this preheating equipment, piping and fittings after the desalters, would reach a value in the order of 45 kgf/cm²a. Higher values could be necessary considering the number of services and heat exchangers and its allowed pressure drop. For them, the pressure class should be 300# or even higher. The metallurgic choice for these equipment should consider the presence of chlorides in an aqueous phase and naphthenic acids. The costs of them would be much higher than the cost of an equipment exposed to naphthenic acids, but not for chlorides in aqueous phase, and pressure class of 150#.

One solution to diminish light hydrocarbons and water content at the preheating train is to use a configuration with a preflash drum or tower and allow them to vaporize under controlled conditions. This will reduce the vapor pressure at the furnace inlet, and eliminate any aqueous phase at the preheating trains after desalters. Using a preflash tower in this case doesn't seem to be the better choice, as for heavy crude oil the content of light hydrocarbons is low for justifying this investment and shifting this material upstream (combustible gas, LPG and light naphtha) tends to affect negatively the vaporization at atmospheric furnace and diesel cut point.

One of the reasons for using a preflash drum is that it is possible to flash and remove a light portion of the crude and entrained water, under controlled conditions, allowing a lower vapor pressure downstream and minimizing the inconvenient effect of shifting material upstream. Higher pressures are also recommended to reduce any foaming at the preflash vessel. Figure 6 presents this configuration. Notice that the vapour phase effluent of the preflash vessel can be added to atmospheric tower via transfer line or at flash zone or even at a higher position at the tower. For each option one has to consider the material and energy balance and the inconvenient of heavy products carry over.

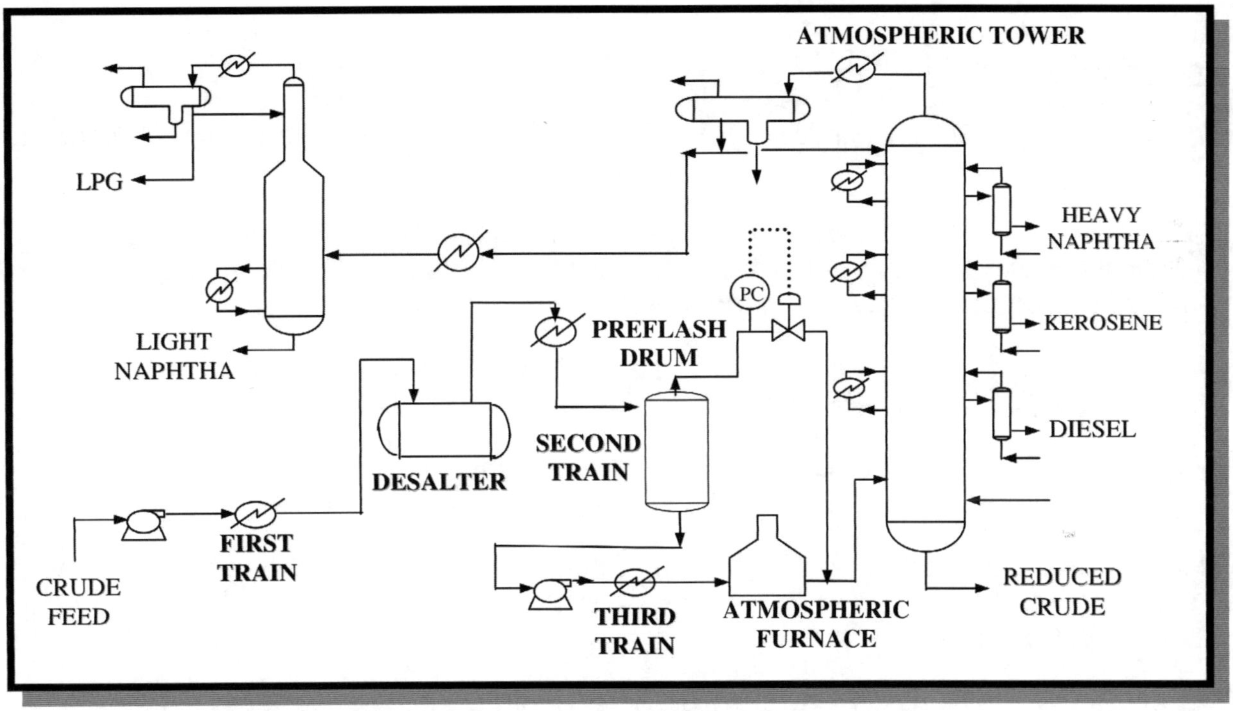

Figure 6: Configuration with Preflash Drum

Conclusions

This paper shows a study of distillation configurations and some aspects related to processing heavy and high naphthenic acids crude oils. The influence of desalted crude oil water carry over associated with very high inlet atmospheric furnace temperature was observed. Simulation results show an important increase in bubble point pressures with the increase in water content at the organic phase. These observations were supported by laboratory determination of bubble point pressures using a PVT cell apparatus.

This paper also presents a case study where even having a heavy feedstock, a configuration with a preflash drum was attractive. Some benefits of this configuration include operational stability of atmospheric furnace and controllability of inlet passes control valves. In addition, reduction in investment cost of preheating train was possible due to minimization at design pressure and a wise metallurgic choice.

References

Andersson, E., (2007) Optimizing Heat Recovery with CPHEs, Petroleum Technology Quarterly, Q1 2007.

Aske, N., Kallevik, H., Sjöblom, J., (2002). Water-in-crude Oil Emulsion Stability Studied by Critical Electric Field Measurements. Correlation to Physico-Chemical Parameters and Near-Infrared Spectroscopy. *J. of Petroleum Sci. And Eng.,* 36, 1-17.

Coutinho, R.C.C., (2005) Estudo da Estabilidade de Emulsões de Água em Petróleos, Dissertação de M.Sc., COPPE/UFRJ, Rio de Janeiro, RJ, Brasil.

Golden, S., (2005) Crude Revamps Using Preflash Drums and Columns, Q&A and Technology Forum.

McLean, J. D., Kilpatrick, P. K. (1997a). Effects of Asphaltene Aggregation in Model Heptane-Toluene Mixtures on Stability of Water-in-oil Emulsions. *J. Colloid Interface Sci.*, 196 (1), 23-34.

McLean, J. D., Kilpatrick, P. K. (1997b). Effects of Asphaltene Solvency on Stability of Water-in-oil Emulsions. *J. Colloid Interface Sci.* 189 (2), 242-253.

Sloley, A.W., (2001) Designing and Revamping Crude Processing Sequences, AIChE Spring National Meeting.

Spangler, R., Hanke, T.A., Varraveto, D., Schoonover, R., (2006) Design of a Crude/Vacuum Unit Revamp for Refinery Flexibility, NPRA Annual Meeting, 19-21 March, AM-06-14.

Sullivan A. P., Kilpatrick P. K. (2002). The Effects of Inorganic Solid Particles on Water and Crude Oil Emulsion Stability. *Ind. Eng. Chem. Res.* 41, 3389-3404.

Waintraub, S., Lemos, A.J.F., Paschoal, L.C.M., Mainenti, M.R.M., (2001) How to Circumvent the Shutdown of a Preflash Tower in an Atmospheric Distillation Unit: an Abnormal Operation, AIChE Spring National Meeting.

Warren, K., (2002) NatcoGroup, The Dual Polarity® Treater – Principles and Operation, Presented at Petrobras, Rio de Janeiro, RJ, Brasil.

Yahyaabadi, R., (2006) Rethink Placement of Preflash Drum in Crude Preheating Train Network, Hydrocarbon Process, August.

Aromatics Extraction in a Single Column to Meet the Demands for Aromatics Production and Benzene Reduction in the Gasoline Pool

Lars Sullivan, Engineering, Services and Equipment, UOP LLC, 25 E. Algonquin Road, Des Plaines, IL 60017-5017

UOP licenses process technology to the refining, gas processing, and petrochemical industries. For over 40 years liquid-liquid extraction, followed by extractive distillation and solvent recovery, has been the preferred method for extraction of aromatics from reformate and pygas streams. In recent years, the refining and petrochemicals industries have embraced extractive distillation alone, followed by solvent recovery, as a way to reduce investment cost. UOP has developed several flowschemes that make use a single column, the ED Stripper, to accomplish the functions formerly performed in the Raindeck Extractor, Raffinate Water Wash and Stripper Columns:

o Separation of aromatics from non-aromatics (Extraction Zone)

o Removal of solvent from the non-aromatics product stream (Rectification Zone)

o Stripping of non-aromatics from aromatics (Stripping Zone)

For some feedstocks, this flow-scheme offers significant savings in capital costs, energy costs, or both. Where applicable, there is also a reduction in plot space requirements and reduced solvent inventory. Two examples of this technology came on-stream in 2006 and two more are expected on-stream by the end of 1Q2007. Examples will be shown to indicate how process considerations influence the design of this column.

Debottlenecking a FCC Gas Plant with Ultra System Limit Consep Trays

Giuseppe Mosca[1], Andrew Hausman[1], Peter Wilkinson[2], Norman Burcher[3], and Nicholas Alsop[4]. (1) Technology, Sulzer Chemtech, Piazza Duca D'Aosta, 12, Milano, 20124, Italy, (2) Shell Global Solutions, PO Box 1, Chester, CH1 3SH, United Kingdom, (3) PreemRaff, Lysekil, Sweden, (4) Preemraff Lysekill, Lysekil, SE-453 81, Sweden

In the middle of 2006 a new Hydrocracker plant was put on stream at Preemraff Lysekil Refinery, Sweden. This affected the feed quality to the FCC Unit and its products yields as well. In particular, the flow rate and composition of the stream going to the debutanizer did change significantly. The intention of Preemraff was to maximize the production of propylene; therefore the catalytic cracking conversion was reset accordingly resulting in a higher feed rate to the debutanizer. This column had been already retrofitted few years back with first generation of High Performance Trays which increased the capacity by 15% over the conventional trays. The column was running at its hydraulic limitation, therefore it could not handle the additional vapor and liquid loadings coming from the latest unit modifications. A new generation of High Performance Trays was required at the stripping section, while multiple downcomer trays could fit in the rectifying section. It was of great importance to minimize the amount of C5 in the top since they damage the catalyst of the down stream Polymerization Unit. It was also important to minimize the C4 in the light cracked naphtha due to the vapor pressure specification. To achieve such challenging targets ConSep trays were selected at the stripping section and HiFi at the rectifying section.

The scope of the paper is to describe the design approach of such difficult revamp projects and report the achieved results.

Spray Section Performance Analysis: an Experimental Approach

André D. Quelhas[1], Antonio V. Castro[2], Glaucia A. Torres[3], Silvia Waintraub[2], Braz Marino Zanutto[4], and Fernado De Carli[4]. (1) RH/UP/ECTAB, Petrobras, Rua Gen. Canabarro, 500, 9° andar - ECTAB, CEP: 20271-900, Rio de Janeiro, Brazil, (2) Cenpes/EB/FCS, Petrobras, Av. Jequitibá, 950, Cidade Universitária, Ilha do Fundão, CEP: 21941-598, Rio de Janeiro, Brazil, (3) AB-RE/TR/TPF, Petrobras, Av. Republica do Chile, 65, CEP: 20031-912, Rio de Janeiro, Brazil, (4) Petrobars/SIX, Parana, Brazil

This work is related with the setup of a new distillation pilot plant. It was designed in order to get information about operational limits and to get insight on physical phenomena within the range of vapor x liquid flow rates. Main focus is to observe the behavior of liquid sprays and wetness of random packing section under several combinations of liquid and gas flow rates. Analysis was made in a qualitative fashion observing liquid and gas stream patterns. Quantitative information was generated and analyzed by digital processing images taken from the spray cone formed. Spray nozzles are directly responsible for geometric shape of the formed liquid spray. Such devices have spread use to promote liquid-gas contact in process operations. Their performance is related to its ability to form stable homogeneous cones of droplets over a region under counter-current gas flow. Thorough studies of such performance are scarce. We aim to share knowledge acquired with tests we designed to evaluate brand new spray nozzles conjugated with its use in a small column resembling a section of a distillation tower. Experimental design consists of a 1590mm height transparent acrylic cylinder. Air is blown from the bottom through a chimney at flow rates in the range 0-680 m3/h. Water flows from the top through spray nozzles at 40-820 kg/h. Both empty and random packed cylinder tests were performed. Pressure drop through nozzles were measured and compared with manufacturer's predictions. Qualitative analysis regarding three parameters was made: fog formation at the top of the cylinder due to droplets carried by ascending air flow, tendency of flooding along packing and air flow instability at the bottom chimney distributor. It was also made an exploratory study in order to quantify information related to spray features by means of image processing. Photographs from a plane parallel to cylinder's top cross section underwent image processing through customized algorithms in order to quantify spray homogeneity along angular sections. In addition, an index was proposed to identify different levels of fog caused by droplets dispersed by air flow. It was possible to identify regions formed by different combinations of the pair liquid x air volumetric flows associated with progressive appearance of fog, air flow instability and flooding and to compare with packing manufacturer's prediction. One spray nozzle showed marked malfunction expressed by unstable spay cone shape, correlated with divergent pressure drop x liquid flow predictions. Quantitative exploratory study showed some non-uniform patterns of liquid distribution normally not noticed by naked eye observation.

230mm
1000mm
360mm
29 8 2006

Smart Plants, Smart Distillation

Jimmy L. Humphrey, Consultant, 10100 Burnet Road, Bldg 133/CEER, Austin, TX 78758, J. Christopher Lewis, Separations Research Program, UT-Austin, 10100 Burnet Road, Bldg 133/CEER, Austin, TX 78758, and Frank Seibert, Separations Research Program, University of Texas at Austin, 10100 Burnet Road, Bldg 133/CEER, Austin, TX 78758.

What are Smart Plants and What is Smart Distillation?

This question will be answered and a practical example will be presented to show how Smart Distillation, based on a recently patented flood predictor model, can be used to maximize column throughput without flooding.

Column Instrumentation Basics: Understanding Accuracy and Sources of Error Part 1

Ruth Sands, Heat, Mass & Momentum Transfer Group, DuPont Engineering Technology, Brandywine 8218, 1007 Market Street, Wilmington, DE 19898

Instrumentation is critical to understanding and troubleshooting all processes. Very few engineers specialize in this field, and most of us learn about instrumentation through experience, myth, and rumor. A good understanding of the various types of instrumentation used on columns can be a valuable tool for engineers evaluating column performance, starting up new towers, or troubleshooting any type of problem. This presentation will give an overview of common types of instruments used for temperature, pressure, level, and flow. A discussion of their accuracy, common installation problems, and troubleshooting examples are also included.

Improving the Efficiency and Reliability of Distillation Model-Predictive Control Projects

Umesh Mathur[1], Mike Morshedi[2], Victor L. Rice[3], Robert D. Rounding[4], Daniel R. Webb[5], and Robert J. Conroy[2]. (1) Techwrite Associates, 15915 Hidden Cove, Houston, TX 77079, (2) Consultant, Houston, TX 77062, (3) APC / RTO, Plant Automation Services, Inc., Suite 600, 16055 Space Center Blvd., Houston, TX 77062, (4) BP America Production Company, 501 Westlake Park Blvd., Houston, TX 77253, (5) Hobbs Fractionation Complex, Ineos, P. O. Box 810, Hobbs, NM 88241

This is an outline of a proposed paper to present a summary / overview of what has been learned in the art of implementing model-predictive control (MPC) projects for large-scale process plants. Traditionally, practitioners have performed extensive and invasive plant tests and used the dynamic response data from the plant to "identify" the required models for the MPC application. In this paper, we will describe the experience gained in four recent projects that have been commissioned using an adaptive multivariable controller. Here, we used first-principles simulation models to eliminate the need for step testing entirely. Such tests are known to be vexatious and expensive, risking violations of process, equipment, safety, and environmental limits.

Earlier in 2003, we had reported experience with use of dynamic simulation for developing the dynamic models required for using an MPC package. These applications were implemented for very slow moving distillation columns (notoriously non-linear and difficult to step-test) with excellent results. Since then, the situation has simplified even further, since the adaptive controller (STAR) used by us in subsequent projects requires the user to provide only the steady-state gains for all input/output variable pairs. The user is not required to specify process dynamics – these are developed adaptively, on-line. Hence, the need for step tests can be eliminated or at least minimized.

The moral of the story is simply this: If you know how to match an Aspen Plus, Hysys, PRO II, or equivalent simulation model to "steady-state" plant data, you can obtain the required steady-state gains quickly by perturbing the model systematically. This enables building an MPC controller in a very straightforward manner. The adaptive controller we used provides one-button CV tuning, which makes commissioning quick and efficient.

All of the MPC projects described in this paper (gas plants and fractionators) were highly successful; two of them were done for BP in the USA, and two the UK. In this paper we will describe the pertinent methodological details about these applications and draw out the common threads, the recommended project execution procedure, along with a list of "dos and don'ts" for ensuring success. Our objective is to show that MPC projects can be done using first-principles knowledge, while minimizing, if not eliminating entirely, the need for empirical dynamic models based on step tests. This should serve as an interesting departure from conventional practice and thus should provide a useful counterpoint to those who hitherto have only seen step tests to build fixed dynamic models.

By focusing on a case-study oriented discussion of what does and does not work, what potential pitfalls exist, and the lessons we have learned, we hope to interest the audience in this approach. Our understanding is that such first-hand accounts are not commonplace in industry, so such discussions should be valuable for those who are practitioners themselves or are responsible for assigning such project work.

Two-Phase-Flow and Stress on Internals during Pressure Relief Events on Distillation Columns: Experimental Investigation and Dynamic Simulation

Daniel Staak, Jens-Uwe Repke, and Günter Wozny. Department of Process Dynamics and Operation, Technical University of Berlin, Straße des 17. Juni 135, Berlin, 10623, Germany

Due to the large content of different substances consisting often of flammable solvents, distillation columns possess a high safety risk potential. Operation failure like the breakdown of different plant components (condenser etc.) can cause the process to leave the field of normal operation and finally cause an emergency pressure relief. The fluid dynamic behavior during the pressure relief event is not well understood. The assumptions in scope of design and operation concerning vapour flow, holdup and pressure drop correlations are made for stationary operations and are not necessarily valid during a pressure relief event. Due to these facts and the unknown column dynamic in this field of operation, still today pressure relief devices especially pressure relief valves (PRV) are designed by guidelines based on experiences and short-cut methods like adaptation of the relief capacity to the overhead vapour rate or the reboiler vapour rate. The consequence of using these methods and to accept the high level of uncertainties is that the relief device has to be oversized to guarantee prevention of the worst case which is a container collapse. On the other hand the oversized relief device is causing a fast depressurization and higher pressure gradients inside the column which implies enhanced vapor flows. Normal tray column internals are designed for a stationary operation load of approximately 20mbar over a stage which can be easily exceeded during a pressure relief event and lead to a destruction of the internals. The total amount of product loss is also increasing due to the oversized relief device. The enhanced vapour flows are causing a transient level swell of the liquid phase which makes it more likely that a two-phase relief flow will occur which should be prevented if possible [1].

The objective is to develop a designing procedure which makes it possible to adapt the cross-section area of the relief valve according to the previous mentioned conditions. The relief area must be as large as necessary to guarantee a mass-flow capacity of the PRV high enough to prevent a container collapse but also as small as possible to avoid a two-phase relief flow and to prevent a destruction of column internals due to the high pressure gradients.

For this new approach methodic experimental analysis were carried out on a pilot scale column with a diameter of 100mm and a height of 4m to receive experimental data concerning response time, outflow (mass and state of phase) and load on column internals during a pressure relief event. The column was additionally equipped with a pressure relief device consisting of a safety valve and a phase separator for determination of the phase fractions during the pressure relief event. Different disturbances of normal operation mode were simulated, which caused an emergency pressure relief event. During the event the level swell of the two-phase region was documented and the relief mass flow and the relief time were determined. Special attention was paid to the phase condition of the relief flow (one-phase flow or two-phase flow).To consider the relevant aspect of stress on internals in the column during pressure relief event, an innovative measurement procedure for online measurement of tray load was developed, commissioned and successfully validated in a testing station. The concept is based on strain gauges, which measure the stress on the tray suspension during operation. By applying a theoretical strain model, conclusions can be drawn concerning the acting force on the column tray during operation. It is a novel approach to measure directly the stresses acting on the material in a distillation column. A comparable strategy was not found in literature. Figure 1 shows the results of the experimental investigation of a pressure relief event in the testing station with a single tray and a gas/vapour flow generated by a blower. During the first 200ms after the opening of the PRV (at 0 s) the pressure drop over the tray is higher than the stationary pressure drop. The maximum pressure drop of approx. 70mbar exceeds the design pressure drop of a standard tray (approx 20mbar) by far. A first dynamic simulation model describing the flow through the PRV was developed and validated with the experimental data (Figure 1).

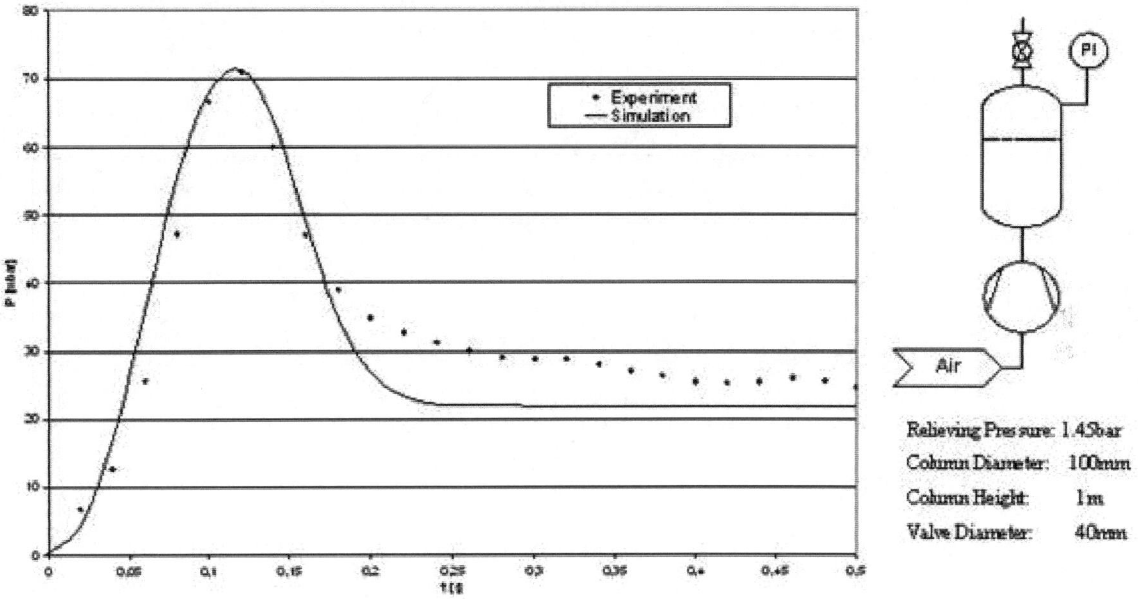

Figure SEQ Figure * ARABIC 1 left: Comparison simulation-experiment, right: testing station

Using the obtained experimental data of the pilot plant concerning pressure gradient, level swell and relief time a rigorous dynamic column model including a PRV was implemented in gProms© to describe the pressure relief event [2]. The model consists of the column including all additional components (condenser, reboiler, controller etc.). The modeling of the PRV was done according to dynamic flow models of PRVs from literature. The PRV model comprises the option of calculating a one-phase or a two-phase relief flow. The opening dynamic of the valve itself (Popen Open Valve, Proportional Valve etc.) is also included in the model. The results show the option of integrating the detailed dynamic pressure relief model in the process simulation for an additional safety analysis of the column. Interaction between different plant components during a pressure relief event can be detected and the appropriate design of the relief capacity for different disturbances of column operation can be checked easily. The model can also be used in the design process. The design of plant components like trays and PRVs can be checked and altered if necessary in an early stage.

In the presentation an overview of the experimental investigation will be given. The online measurement procedure for measuring stress on column internals will be presented. The key features of the dynamic model will be explained and a discussion of simulation and experimental data will be done.

The authors would like to thank the Nafög-Foundation for sponsoring this research project.

[1] Ü.Can, M. Jimoh, J. Steinbach, G. Wozny, *Simulation and experimental analysis of operational failures in a distillation column,* Separation and Purification Technology 29 (2002) 163-170

[2] D. Staak, J.-U. Repke, G. Wozny: *Outflow conditions and strains on internals during pressure relief events on distillation columns*, Pres Conference 2006, Prag

Improve Column Control by Modifications to Temperature Control Strategies

Ruth Sands, Ken Jackson, and **Jim England**. Heat, Mass & Momentum Transfer Group, DuPont Engineering Technology, Brandywine 8218, 1007 Market Street, Wilmington, DE 19898

As rates are increased on existing towers, quality and stability issues may surface even though the column internals can theoretically handle the increased production. Sometimes these issues can be solved by implementing a modification to a simple temperature control strategy. Temperature averaging is a very stable control method that uses two or more measured temperatures to compute an average, which is then used as the control point. This method is especially useful for towers that have a large temperature difference between top and bottom. Several examples of improved stability with regard to rate changes and process upsets will be given. Pressure compensated temperature control is another variation of the standard temperature controller that can be used to alleviate quality and capacity issues. Two examples of improved control will be presented, along with a guide on how to implement these strategies.

Control of a Batch Distillation Column

Hongbin Dong and **Jim Riggs**. Texas Tech University, Dept of Chemcial Engineering, Lubbock, TX 79410

The primary challenge associated with the control of a batch distillation process is that process gain for the overhead product decreases continuously as the overhead product is removed. An analysis of the process gain of a rectifying section of a column using the Smoker equation assuming a binary separation shows that the overhead gain varies somewhat linearly with the composition of the light key in the base of the column. Based on this generalized analysis, an approach for scheduling the gain of the overhead composition controller for a batch distillation column was developed. The process gain at the beginning of the batch and at the end of the batch are measured experimentally. These results are used to estimate the current process gain using the online estimate of the composition of the light key in the column bottom and assuming that the process gain varies linearly with the composition of the light key in the column bottom. The product composition and flow rate are used online to continuously estimate the composition of the light key in the column base. The control configuration used in this study is applied by cascading the overhead composition controller to a tray temperature controller for a selected tray near the upper portion of the column. The gain of the composition controller is adjusted by linearly adjusting the process gain during the production phase while a fixed gain temperature controller is used.

This approach for batch distillation control was tested using a detailed dynamic simulation of a multi-component batch distillation column. The batch column contained ethane, propane, iso-butane, n-butane and n-pentane. The initially an ethane and propane product was produced. Startup of the column under total reflux conditions was consider followed by transition to the production phase for the propane and ethane product. At the end of the ethane and propane production phase, a transition to the production phase for iso-butane was simulated. Finally, the complete production phase for the iso-butane product was simulated. In spite of the fact that the process gain for this system changes by over an order of magnitude, the proposed gain scheduled controller demonstrated very tight composition control for both batch production phases.

Column Instrumentation Basics: Understanding Accuracy and Sources of Error Part 2

Ruth Sands, Heat, Mass & Momentum Transfer Group, DuPont Engineering Technology, Brandywine 8218, 1007 Market Street, Wilmington, DE 19898

Instrumentation is critical to understanding and troubleshooting all processes. Very few engineers specialize in this field, and most of us learn about instrumentation through experience, myth, and rumor. A good understanding of the various types of instrumentation used on columns can be a valuable tool for engineers evaluating column performance, starting up new towers, or troubleshooting any type of problem. This presentation will give an overview of common types of instruments used for temperature, pressure, level, and flow. A discussion of their accuracy, common installation problems, and troubleshooting examples are also included

Energy Use in Distillation Operation: Nonlinear Economic Effects

Douglas C. White, Emerson Process Management, 2119 Albans, Houston, TX 77005-1517

Distillation operations are major consumers of energy, by some estimates comprising forty percent of the energy usage in the refining and chemicals industry. Obtaining the maximum energy efficiency from this unit operation is obviously very important. For many distillation columns there is a tradeoff in operation between energy usage and product recovery and setting the proper usage target involves a calculation of the economic tradeoff between these two factors. However, distillation is a non-linear process and normal economic valuations add more non-linearities to the economic objective functions. In addition, the normal product quality variability observed leads to requirements for statistical evaluation. Hence, calculation of the correct target can be complicated. In this paper, these non-linear economic effects are reviewed and techniques to calculate the correct usage targets presented.

The Effect of Hole Size on Sieve Tray Performance

Mitch Sakata, T. Yanagi, D.W. King, Tony Cai, and **G.X. Chen**. Fractionation Research, Inc., P.O. Box 2108, Stillwater, OK 74076

The effect of hole size on sieve tray performance was systematically studied in a 1.22 m distillation column with various tray configurations. The diameters of the sieve holes covered a range from 3.2 to 66.7 mm. The tests were conducted with three different hydrocarbon systems, cyclo-hexane/n-heptane, iso-butane/n-butane, and o/p-xylene at pressures of 0.02 to 27.6 bar. The performance of the sieve trays with different hole sizes was determined and compared. It is found that hole size affects the mass transfer efficiency and capacity. The results also show a significant effect of the hole size on tray pressure drop and weeping. A fundamental analysis is given to explain the effect of hole size on the sieve tray performance.

High Capacity Tray Revamp of a C2 Splitter

Daniel R. Summers, Sulzer, 4019 South Jackson Avenue, Tulsa, OK 74107, Andre Bernard, NOVA Chemicals (Canada) Ltd., 785 Petrolia Line, Corunna, ON N0N 1G0, Canada, and Waldo E. De Villiers, Shell Global Solutions (US) Inc., 3333 Highway 6 South, Houston, TX 77082.

The NOVA Chemicals (Canada) Ltd. complex located in Sarnia, Ontario, Canada was recently revamped in 2005 to improve the C2 Splitter tower performance. After careful examination of the existing design, desired operation, and future capacity, HiFi™ trays were chosen to replace the old trays. An unusual approach was taken to achieve desired performance. Trays were replaced at larger tray spacings with fewer trays. An increase in tray efficiency was expected and required to achieve the overall performance that was desired after the revamp. After a successful installation, the unit was restarted in late 2005 and desired performance was achieved.

This paper will present the actual operating information for this tower so that others may evaluate the performance, capacity and efficiency of the HiFi tray in C2 Splitter Service.

Increasing the Capacity and Lowering the Energy Consumption of a Paraffin Fractionation Tower

Greg Spencer, Koch-Glitsch, UK, Stoke-on-Trent, United Kingdom and Izak Nieuwoudt, Koch-Glitsch, 4411 E 37th Street N, Wichita, KS 67220.

A UK refiner has been operating a paraffin fractionation tower equipped with modern high-performance trays. They were experiencing operational difficulties and wanted to do a quick shut-down to fix the problems. Due to the severity of the problems any new equipment had to be on site within 4 weeks. In analyzing the performance of the tower Koch-Glitsch engineers realized that not only could the problems be fixed, but the capacity could be increased AND the energy consumption reduced. The equipment were designed for quick installation and delivered within 4 weeks. The shut-down time needed to install the equipment was very short compared to what was anticipated. The tower was started up without any problems and produced a better product while using less energy. The capacity of the new internals exceeded the capacity of the auxiliary equipment. At the next shut-down the axiliary equipment will be de-bottlenecked to enable the refiner to utilize the full capacity of the new internals. This paper describes the details of this revamp.

The Effect of Open Area on Sieve Tray Performance

Mitch Sakata, T. Yanagi, A. Shariat, Tony Cai, and **G.X. Chen**. Fractionation Research, Inc., P.O. Box 2108, Stillwater, OK 74076

The studies described in this paper were made to understand the effect of open area on sieve tray mass transfer and hydraulic performance. Tests were conducted in the 1.22 m and 2.44 m diameter sections of Fractionation Research, Inc. (FRI) experimental distillation unit. The sieve trays were tested over a wide range of open areas from 5 to 30 percent. Three different hydrocarbon systems, cyclo-hexane/n-heptane, iso-butane/n-butane, and o/p-xylene, were used at pressures of 0.02 to 27.6 bar. Experimental results, including tray capacity, efficiency, pressure drop, and weeping, are presented and discussed. The performance of sieve trays with different open area was determined and compared. Explanations of the test results are given based on the fundamental analysis.

Hydraulic Investigations to Increase the Capacity of Trays in Distillation Columns

M. Jödecke, T. Friese and G. Schuch
GCT/C – L540, BASF AG, 67056 Ludwigshafen, Germany
email: michael.joedecke@basf.com

Abstract

Experimental investigations of elements of high capacity trays were performed in a 0.5 m air-water column. The investigations include hydraulic measurements of new valve types, of truncated downcomers and of de-entrainment devices. The influence of support beam orientation at the capacity limit of a tray and special types of high capacity trays were investigated as well.

The experimental results for the truncated downcomers underline a wide range of design. De-entrainment devices can significantly increase the capacity of a tray, but they are only useful only for modestly liquid loads. The optimal void fraction of the de-entrainment device has still to be found. The investigations of the support beam orientation revealed, that there's no influence of the flooding point F factor of the tray when the support beam is oriented lengthwise, but an influence is observed for crosswise oriented support beams. For modestly liquid loads the crosswise oriented support beam increases the flooding point of the tray, the contrary effect was noticed for higher loads.

Introduction

Trays are often used internals in distillation columns. Since several years the producers of column internals offer different devices to increase the capacity and/or the efficiency of distillation trays. There's a lack of independent investigations of these types of trays.

It is possible to increase the capacity or the efficiency of trays in different ways:

- new geometric forms of valves on the trays, e. g. fixed valves, especially made for high gas loads,
- the use of de-entraiment devices above the tray to reduce entrainment of liquid to the next tray,
- increase of the active area of the tray by using reduced or truncated downcomers and
- special constructions of trays with certain devices to separate gas and liquid e. g. by using centrifugal forces instead of gravity forces.

The combination of these types of devices is also possible. Trays, which were changed in this way, also are designated as high capacity trays.

In a research project elements of high capacity trays were investigated experimental in an air-water column. New types of valves special made for high gas throughput were investigated, e. g. BiFrac, VG0, ProValve, MVG, SVG and BDH valves. These investigations were completed by experiments to truncated downcomers, to de-entrainment devices and some special

constructions like the CoFlowTray of the Chinese company Tianjin Corp. Also hydraulic investigations to the influence of the support beam orientation at the flooding point F factor of the trays were performed.
The results of the investigations to truncated downcomers, de-entrainment elements and influence of the support beam orientation will be presented.

Experimental apparatous and procedure

The experiments were performed in a 0.5 m air-water column under environment pressure and environment temperature conditions. The tray space was 0.41 m.
The liquid flow rate in the column can be adjusted by pump until 100 $m^3/(m^2h)$. A blower produces the gas stream; the F factor in the column can be subject to variation until 8 $Pa^{0.5}$.
Usually three trays can be measured simultaneously. The equipment is able to estimated pressure drop, entrainment rate and weeping rate of the trays. The results can be used to estimate e. g. the flooding point of the trays.
Electric pressure transducers measure the pressure drop. The equipment is able to measure the pressure drop of the three single trays as well as the pressure drop of the whole column. Entrainment and weeping rate can be estimated volumetric.

Experimental results

Investigations of truncated downcomers

Truncated downcomers are used to increase the active area of a tray. The win of active area is the area of the saved downcomer from the tray above.
The open area of the downcomer outlet limits the design of a truncated downcomer. Is the downcomer outlet to small, the downcomer floods to early and limits the capacity of the tray. Is the open area to big, the vapour phase is bypassing the column trough the downcomers. That reduce the contact between liquid and vapour phase and reduce the efficiency of the column.
The experiments with the truncated downcomers were carried out with sieve trays. The hydraulic design of the trays were:

- spec. active area 0.75
- spec. downcomer area 0.075
- 680 sieve holes with 6 mm in diameter

To investigate the influence of the downcomer outlet area several downcomers with an open area of 2.1, 7.8, 45 and 100 % were produced. The open area was created by holes of 10 mm in diameter (4, 15 and 45 holes).
On of the first result was, that it's not possible to combine a truncated downcomer with a sieve tray. The liquid stream out of the downcomer goes directly trough the holes of the sieve tray and falls in the downcomer of the tray below. The result is less cross flow over the tray. To get a cross flow over the tray the sieve holes under the downcomer were covered. That's why the investigations were not useful to determine the win in capacity of the tray by using a truncated downcomer, but useful to understand the behaviour of the downcomer. Height of liquid in the downcomer and bypassing of gas trough the downcomer were estimated.

Figure 1 shows some of the experimental results of these investigations. The figure includes the results for downcomer flooding of a downcomer with 4 holes (2.1 % open outlet area) compared with calculation results for downcomer flooding. The available models are not able to describe the downcomer flooding of the truncated downcomer. But the difference can be neglected. Only small changes in the open area in the calculation lead to better description of the experimental results. With the available equipment it was not possible to flood the downcomer with an open area of 24 % (45 holes).

Only for the downcomer with 100% open area a bypassing of gas was observed. That proves a wide range of the design of truncated downcomers.

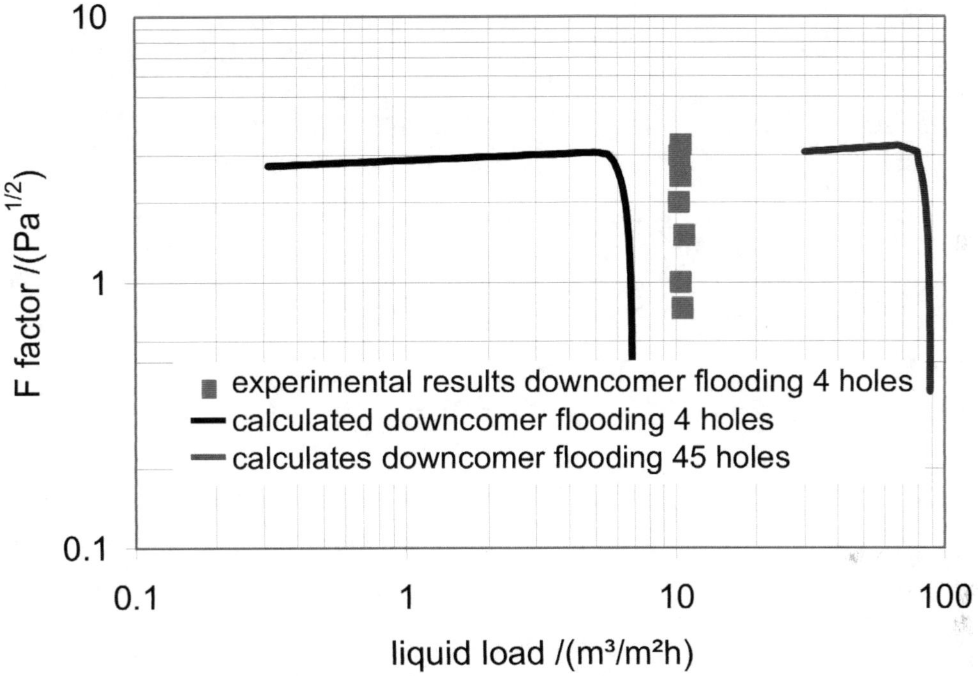

Figure 1. Comparison of experimental and calculated results for downcomer flooding of truncated downcomers with different open outlet areas

Investigations of de-entrainment devices

De-entrainment devices at the trays are used to reduce the entrainment of liquid to the tray above. This increases the capacity of the tray without lost of efficiency. As de-entrainment elements random and structured packings are used. These investigations were done with Pall rings (50 x 50) packed at the active area of sieve trays. The hydraulic design of the modified trays were:

- spec. active tray area 0.75
- spec. downcomer area 0.075
- 183 holes with 12 mm in diameter

Figure 2 shows a typical experimental result for a liquid load of 8 $m^3/(m^2h)$. The weeping rate of the tray is decreasing with increasing gas load. Pressure drop and entrainment increase with increasing gas load. The results show a significant entrainment from F factors higher than 4 $Pa^{0.5}$. That proves a strongly increased working area of the modified sieve tray. But the pressure drop of the tray in this range of gas load is higher than 20 mbar. Because of what explained above, this can cause problems with the mechanical stability of the tray in particular for columns with larger diameters.

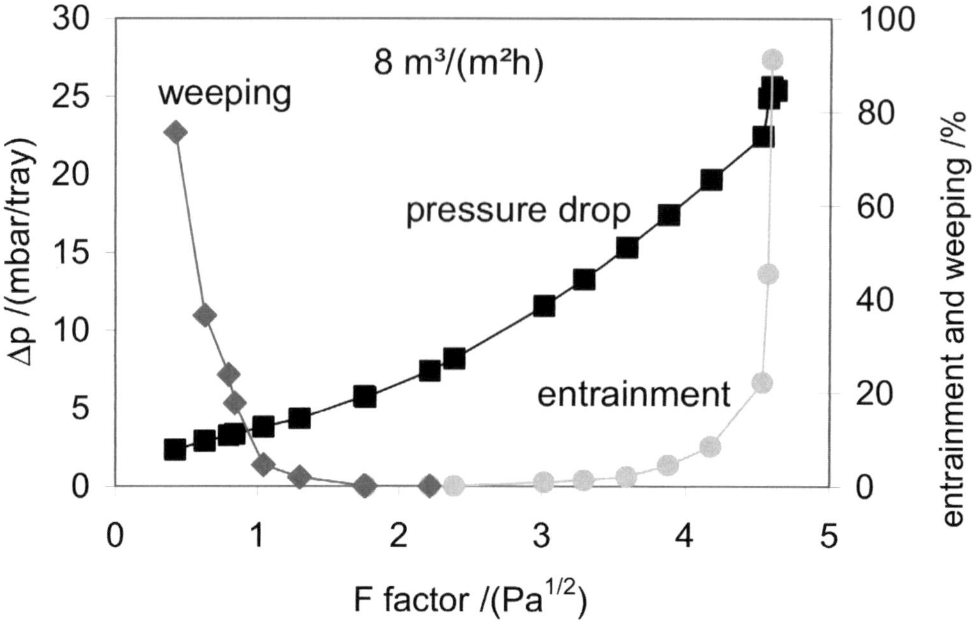

Figure 2. Experimental results for a sieve tray with Pall rings at constant liquid load of 8 $m^3/(m^2h)$

Figure 3 summarize the experimental results for the flooding point F factor of the modified sieve tray and include comparisons with calculation results of flooding point F factors for Pall ring packings and for the sieve tray without modifications. The results of the experiments indicate a significant increase of the flooding point of the modified sieve tray for liquid loads smaller than 10 $m^3/(m^2h)$. With increasing liquid load the curve is falling. For high liquid loads the flooding point F factor of the sieve tray with Pall rings is lower than for the sieve tray without modifications. The flooding point curves for the Pall ring packings were calculated with a reduced diameter (D = 0.436 m) to regard the fact, that only the active area of the tray is covered with Pall rings. For low liquid loads the model is able to describe the jet flood curve of the modified sieve tray. For higher liquid loads the model overestimate the flooding point of the sieve tray with the Pall ring packing. The calculated flooding points for smaller Pall rings (Pall ring 25 x 25) highlight that the use of smaller rings reduce the possible capacity of the modified sieve tray. There must be an optimum for the size of the rings or the void fraction of the packing to get the best possible capacity increase of the tray.

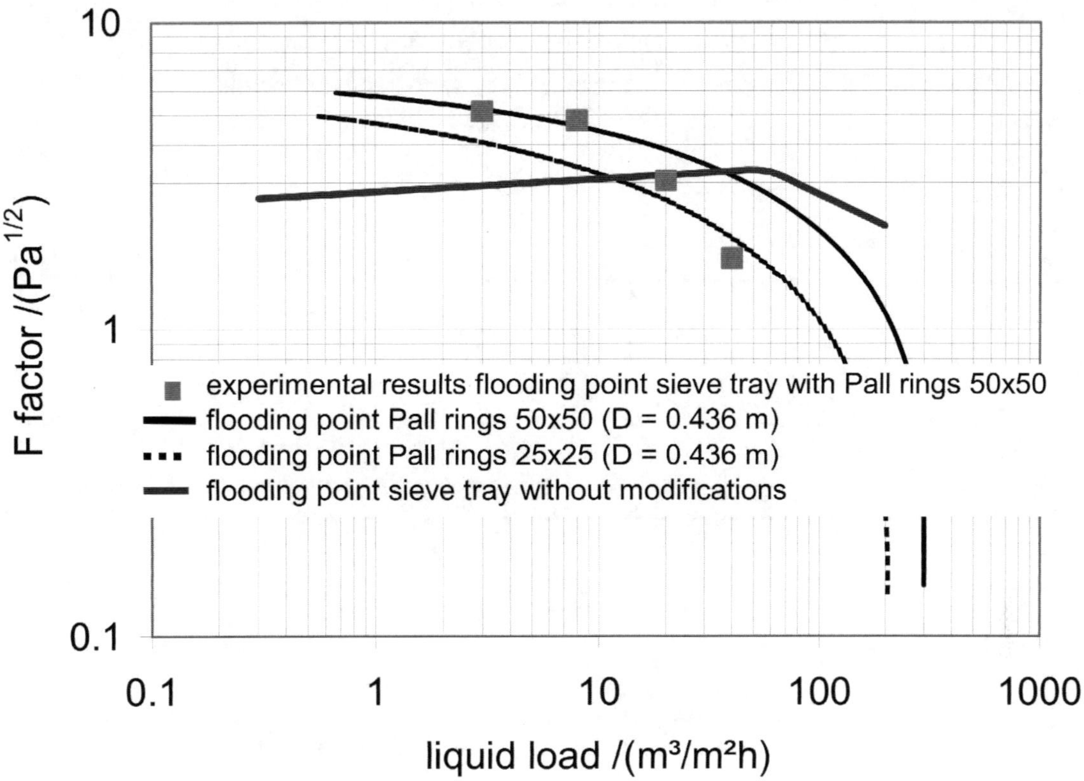

Figure 3. Experimental and calculated results of flooding point F factors for a sieve tray with Pall ring packing

Investigation of support beam orientation

Several investigations with different oriented support beams have been executed in order to understand the influence of the support beam orientation at the capacity of trays. The support beams were located under the tray lengthwise and crosswise to the liquid flow direction on the tray. The investigations were performed with MVG valve trays of SULZER. The hydraulic design of the trays were:

- spec. active area 0.75
- spec. downcomer area 0.075
- 49 MVG valves with 6 mm valve height

Figure 4 compares the results for the flooding point and 10% of entrainment for the MVG valve tray with lengthwise support beams and without support beams. There's no significant difference between the experimental results for both facilities.

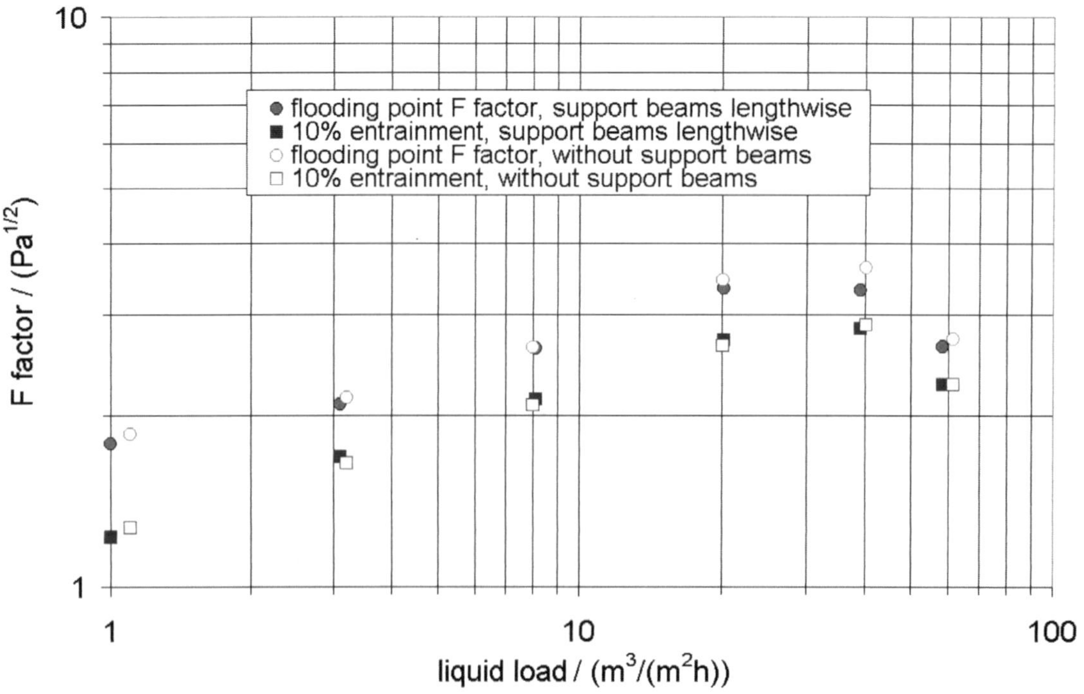

Figure 4. Comparison of experimental results for flooding point F factor and 10% of entrainment for a MVG valve tray with and without support beam oriented lengthwise

Figure 5 compares the results for flooding point and 10% of entrainment for a MVG valve tray with and without crosswise oriented support beams. For modestly liquid loads a crosswise oriented support beam increases the capacity of the tray by nearly 10 percent. The support beam has the function of a de-entrainment device. For higher liquid load the effect of the crosswise oriented support beam turns round and reduce the capacity of the tray.

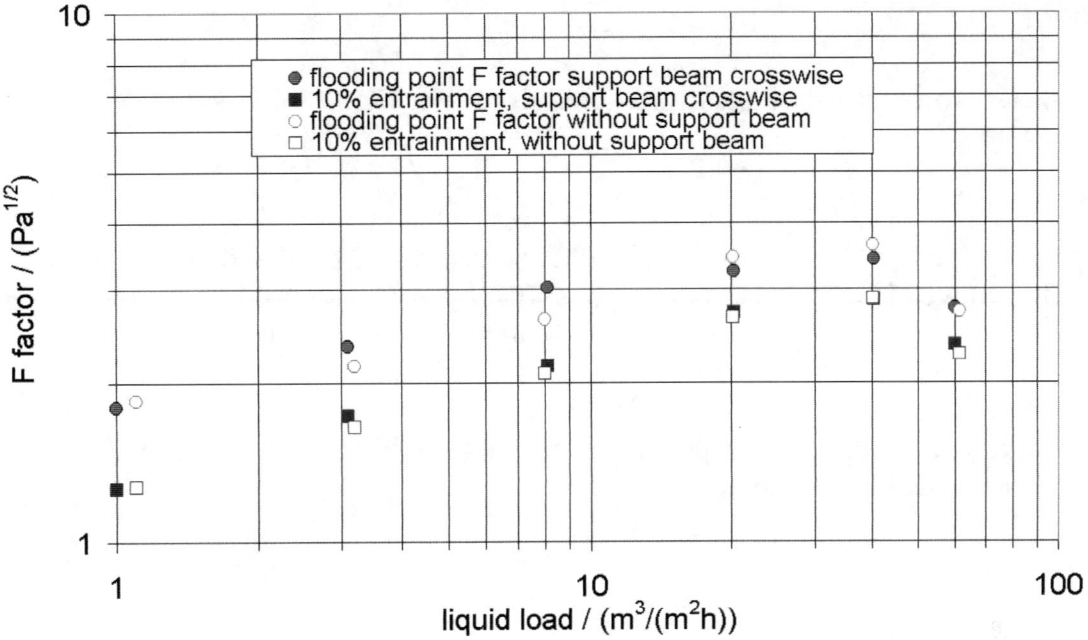

Figure 5. Comparison of experimental results for flooding point F factor and 10% of entrainment for a MVG valve tray with and without support beam oriented crosswise

Summary, Open Questions

Several experimental investigations to elements of high capacity trays were performed in a 0.5 m air-water column. The experimental results helps to understand behaviour of these elements in distillation columns.
The experiments for truncated downcomers point out a wide range of design of these types of downcomers. Bypassing of gas seems to be no problem. The outlet area of the tray can be designed with the common models. Truncated downcomers cannot be used in combination with sieve trays.
De-entrainment devices at trays can increase the capacity of the tray significantly for modestly liquid loads. The choice of the right de-entrainment element is important. The optimal void fraction of the de-entrainment devices must be estimated.
The orientation of support beams impacts the capacity of a tray. Crosswise oriented support beams increase the flooding point of a tray for modestly liquid loads and reduce this point for higher liquid loads. The orientation of the support beams must be considered in the design of the tray.

Parallel Flow Trays

Ping Xu, Brian Nowak, Kevin Richardson, and **Bob Miller**. UOP LLC, 175 E. Park Drive, Tonawanda, NY 14051-0986

It is well known that parallel liquid flow on adjacent distillation trays eliminates concentration pinch and improves tray efficiency. The best-known distillation tray with a parallel flow arrangement is the UOP Parallel Flow Slotted Sieve tray used primarily in EB/Styrene and POSM applications. UOP has recently incorporated parallel flow into its high capacity Multiple Downcomer tray to enhance mass transfer efficiency. This tray – the PFMD tray – retains the high capacity characteristics of the MD family of trays with a tray efficiency similar to or higher than conventional trays. The PFMD tray is expected to be used in both low and high operating pressure applications to increase column capacity near to ECMD capacity while maintaining a high tray efficiency.

Improving Efficiency of Multiple Downcomer Tray Performance by Various Tray-structure Modification

Kejian Yao[*], *Lianghua Wang* , *Lingyu Zhu and Xiaomei Yu*

State Key Laboratory Breeding Base of Green Chemistry Synthesis Technology

College of Chemical Engineering & Materials Science

Zhejiang University of Technology

Hangzhou, 310032

P. R. China

Statement:

This paper is presented in the Distillation Topical Conference of the 2007 Spring AIChE Meeting, Houston, TX, April 22-26, 2007

[*] Corresponding author. Tel: 86-571-88320706; Fax: 86-571-88320712; E-mail: yaokj@zjut.edu.cn

Improving Efficiency of Multiple Downcomer Tray Performance by Various Tray-structure Modification

Kejian Yao, Lianghua Wang , Lingyu Zhu and Xiaomei Yu*
(State Key Laboratory Breeding Base of Green Chemistry Synthesis Technology, College of Chemical Engineering & Materials Science, Zhejiang University of Technology, Hangzhou, P. R. China, 310032)

ABSTRACT: Promotions of multiple downcomer trays were studied and discussed. Directing baffles were added on tray deck to reduce the liquid back mixing, resulting in uniform liquid distribution. A bed of structured packing was hung under the tray deck to capture and reduce the entrainment. Fixed valves with downward inclined margins were employed as bubble units to reduce the entrainment and weeping and to increase the chance of gas-liquid contacting. The experimental results showed that when the F-factor increased to 2.0, the entrainment fraction of the tray with the packing was 0.003 kg water/kg air, obviously lower than that of the tray without the packing, 0.014 kg water/kg air. This indicated that the combined packing technology could reduce the entrainment significantly. The experimental results also showed that when fixed valves with downward inclined margins were employed as bubble units, the F_0 factor of entrainment flooding point was larger than sieve tray and V1 valve tray. Anti-dumping units were applied to prevent the liquid from dumping through orifices. The experiment with ethanol/water system showed that the DJ-3 tray efficiency was 10%-15% higher than that of the conventional multiple downcomer tray. The experiment with cyclohexane/N-heptane system showed that the DJ-5 tray efficiency was 10%~ 20% higher than that of conventional multiple tray in all operating load and as large as V1 valve tray in middle and large operating load.

KEYWORDS: DJ-3 tray, DJ-5 tray, multiple downcomer tray, tray efficiency.

1. Introduction

The conventional multiple downcomer trays consist of sections of perforated tray decks separated by parallel trough-like downcomers. The downcomers of sequent trays are arranged perpendicularly[1] and suspended on the tray with the bottom terminated in the vapor space above the froth of the tray below. The liquid spouts through the holes at the bottom of the downcomers. Consequently, receiving pan is no longer required on the tray deck. The absence of this receiving pan enables an additional 10%-15% of the column cross-sectional area to be useful for the vapor traffic. Meanwhile, the whole perimeter of the trough-like downcomer is fully used as the outlet weir, resulting in a lower liquid load on the weir and a higher capacity.

On the other hand, the conventional multiple downcomer tray features a relatively lower efficiency than the tray with regular downcomers mainly for the mal-distribution of liquid on the tray deck. This results in requirement of extra column or trays in revamping process separation [2,3]. As commonly known, the ideal liquid flow is the plug-flow on a well-designed tray. On segmental downcomer trays, the downcomers of sequent trays are parallel and the approximate plug-flow can be obtained between the inlet weir and the outlet weir except for the segmental region. On the multiple downcomer tray deck, however, the plug-flow can no longer be guaranteed when the downcomers of sequent trays are settled perpendicularly to each other.

Another factor causing the low efficiency of multiple downcomer trays is the dumping on the perforated receiving area of the tray, the so-called impact dumping. As a result, the liquid spouting from the exit holes of the downcomer will probably spurt directly onto the bubble area and a portion of the liquid will penetrate the orifices and dump directly onto the downcomers of the next tray below. This type of liquid shortcut results in less mass transfer and consequently the lower tray efficiency.

To ameliorate the efficiency of the multiple downcomer trays, modifications of the tray structure are presented in this paper. The tray efficiency enhancement devices are studied and two improved multiple downcomer tray, DJ-3 tray and DJ-5 tray, are developed. Experimental results are also presented.

* Corresponding author. Tel: 86-571-88320706; Fax: 86-571-88320712; E-mail: yaokj@zjut.edu.cn

2. Experimental setup

The experiments conducted in this project include five parts. Firstly, the liquid residence time distribution (RTD) on the tray deck was tested with fiber-optic probes to investigate the effect of adding directing baffles on the liquid profile of the multiple downcomer tray. These experiments were conducted in a 1200 mm diameter simulator with air/water system. The simulator was equipped with two identical multiple downcomer trays each with a trough-like downcomer. The test tray is referred as tray #1 in this paper. The sketch of test tray is shown in Figure 1.a, together with nine other test trays described in later paragraph. The directing baffles, which are in arc shape, are made of 100mm×150mm rectangular steel plates. The detailed dimensions of tray #1 are listed in Table 1.

Secondly, experiments were conducted to test effect of adding structured packing on the entrainment in a 300 mm diameter and 350 mm tray space air/water simulator. A thin layer of structured packing was hung and fixed by bolts under the tray deck with the same shape and scale as the tray deck. The structure packing used in the experiment is Mellapack. The height of the thin layer is 100mm. As shown in Figure 2, three identical multiple downcomer trays were installed in the simulator. The downcomers of the first tray from the top were used as the liquid distributor of the second tray. An entrainment catcher mounted on the top tray captured the liquid droplet entrained by the ascending vapor from the second tray. The entrainment captured was periodically weighed. The sketch of the test tray, numbered as test tray #2, is illustrated in Figure 1.b. Dimensions of the tray tested in this experiment are also presented in Table 1.

Thirdly, experiments were carried out to measure the effect of fixed valve as bubble unit on entrainment in a 600 mm inner diameter and 600 mm tray space transparent plexiglass column with air-water system. The schematic diagram of this experiment is also illustrated in Figure 2. Fig.3 is a perspective view showing a group of fixed valves with downward inclined margins on tray deck. The method of measuring entrainment is same as before. The sketch of the test tray, numbered as test tray #3, is illustrated in Figure 1.c. Dimensions of the tray tested in this experiment are also presented in Table 1.

Fourthly, the effect of adding anti-dumping units on dumping was tested in a 500 mm diameter and 600 mm tray space air/water simulator. The schematic diagram of this experiment is also illustrated in Figure 2. As shown in the figure, the top multiple downcomer tray was used as a liquid distributor and the dumping liquid from the middle tray was collected in the container under the bottom tray. The dumping liquid was periodically introduced to the cylinder and weighed. The sketch of the test tray, referred as test tray #4, and the location of the anti-dumping units are illustrated in Figure 1.d. Dimensions are listed in Table 1.

Lastly, the experiments were conducted to determine the tray efficiency in a 300 mm diameter and 350mm tray space ethanol/water distillation column consisting five identical trays or in a 600mm diameter and 580mm tray space cyclohexane/N-Heptane distillation column consisting of six identical trays. Two columns were made of stainless steel. The first and the second trays from the top were equipped with thermocouples and sampling points. The third sampling point was equipped on the bottom tray. A schematic diagram of the experimental apparatus is shown in Figure 4. The column, mounted on a vessel with a serpentuator heater in it, was equipped with a total condenser and total reflux. To test the effect of adding directing baffles and combined packing on the tray efficiency, the test tray #5, as shown in Figure 1(e), equipped with the baffles and the packing, was used. In addition, two arch downcomer valve trays, referred as test tray #6 and #7, were also used for efficiency comparison. The test tray #8, a conventional multiple downcomer tray added with the anti-dumping units, as shown in Figure 1(d), was also designed to test the effect of adding anti-dumping units on the tray efficiency. To test the efficiency of DJ-5 tray, the test tray #9, a multiple downcomer fixed valve tray added with directing baffles and anti-dumping units was used. The test tray #9 is illustrated in Figure 1(f). In addition, a conventional multiple downcomer tray #10 was used for efficiency comparison. Detailed dimensions of those test trays are shown in Table 1.

3. Results and discussion

The efficiency enhancement devices tested in this study include the directing baffles[4,5], the combined packing technology[6,7], fixed valve as bubble unit[8] and the anti-dumping units[9]. The detailed results are presented in the following.

3.1. Effects of adding directing baffles on liquid distribution

Figure 5.a is a contour plot of the liquid RTD on a quarter of the test tray #1 without directing baffles, in which the points in each curve are with the same mean residence time represented by an attached number in the unit of second. The results were obtained at the liquid rate of $8m^3/m^2{\cdot}h$ and the vapor rate of $1600m^3/m^2{\cdot}h$. It is shown that the residence time of the liquid in the center of the tray is larger than that along the wall, indicating a slow rate pool in the center caused by the outlet weir. The results prove that the flow pattern cannot be considered as plug-flow in this type of tray.

To keep the liquid flowing be uniform in a moderate rate, baffles were tentatively bolted on different locations of the tray deck to prevent shortcut flow. Experiments were conducted to investigate the effect of the directing baffles on the liquid RTD. The contour plot of the liquid RTD on a quarter of test tray #1 with the directing baffles is illustrated in Figure 4.b. The results were obtained at the same liquid and vapor rates as described above. It is shown that, after the directing baffles were installed, the liquid along the wall was forced to move slowly and the liquid in the center was pushed to move fast. Owing to the directing effect of the baffles, the RTD of the liquid along the wall turned to be larger and the liquid in the center smaller and the RTD distribution, as a consequence, became uniform.

3.2. Effect of adding combined packing on entrainment

The combined packing technology was also used to improve the performance of the tray. Experiments were conducted to investigate the effect of adding the combined packing on entrainment. All parallel experiments with and without the packing were carried out in the 300mm diameter air/water simulator with test tray #2. The entrainment was measured above the test tray with and without combined packing, respectively. Both results are shown in Figure 6, where the entrainment fraction was defined as the ratio of the mass rate of liquid entrained by ascending gas (M_{Le}) over the mass rate of ascending gas (M_G),

$$e_v = \frac{M_{Le}}{M_G} \qquad (1)$$

The liquid rate was fixed at $8.5\text{m}^3/\text{m}^2\cdot\text{h}$ and the F-factor (F) varied from 0.7 to $2.7\text{m/s}\cdot(\text{kg/m}^3)^{1/2}$, where the F-factor is defined as the square root of the gas kinetic energy based on cross sectional area of the column,

$$F = \sqrt{\rho_G \left(\frac{V_G}{\frac{\pi}{4}D^2}\right)^2} \qquad (2)$$

It can be seen from Figure 6 that the entrainment above the conventional multiple downcomer tray without the packing is obviously higher than that with the packing. At low vapor rate, the entrainments on both trays can be ignored; and it increases with the increase of vapor rate. At high vapor rate, considerable entrainments can be observed on the tray without the packing. When the F-factor increases to 2.0, the entrainment fraction of the tray without the packing is 0.014 kg water/kg air, which is obviously larger than that of the tray with the packing, 0.003 kg water/kg air. This indicates that the combined packing technology reduces the entrainment significantly. It is due to the packing, which was settled in the disengagement space between the adjacent trays, captured the ascending liquid droplet and hence reduced the entrainment at the high vapor load.

3.3. Effect of fixed valve with downward inclined margin as bubble unit on entrainment

Entrainment changes with gas/liquid flow rate in a multiple downcomer tray with fixed valve were same as that in a sieve or valve tray shown in Fig.7. But, the fixed valves restrained the entrainment. The gas ascended upward through valve apertures, and then thrust into the cross-flowing liquid in a horizontal direction due to the presence of the valve cover mounted above the aperture, rather than vertically in sieve tray, which reduced entrainments. Many researchers showed that excessive entrainment reduced tray efficiency and quickly induced entrainment flooding. According to usual design rules, the upper limit of entrainment should be less than 0.1 kgLiquid· (kgGas) $^{-1}$. The F-factor is named as entrainment flooding F-factor (F_{flooding}) when the entrainment is equal to 0.1 kgLiquid· (kgGas) $^{-1}$. The experimental results showed that F_{flooding} was 2.45 $\text{kg}^{0.5}\cdot\text{m}^{-0.5}\cdot\text{s}^{-1}$ at liquid flow rate of 2.1 $\text{kg}^{0.5}\cdot\text{m}^{-0.5}\cdot\text{s}^{-1}$ at 20 $\text{m}^3\cdot\text{m}^{-2}\cdot\text{h}^{-1}$ and liquid flow rate of 80 $\text{m}^3\cdot\text{m}^{-2}\cdot\text{h}^{-1}$ in present multiple downcomer tray employed fixed valve as bubble unit. According to Hunt's correlation, F_{flooding} is 2.2 $\text{kg}^{0.5}\cdot\text{m}^{-0.5}\cdot\text{s}^{-1}$ at liquid flow rate of 20 $\text{m}^3\cdot\text{m}^{-2}\cdot\text{h}^{-1}$ and 1.7 $\text{kg}^{0.5}\cdot\text{m}^{-0.5}\cdot\text{s}^{-1}$ at liquid flow rate of 80 $\text{m}^3\cdot\text{m}^{-2}\cdot\text{h}^{-1}$ in a sieve tray column with same tray spacing and same relative open area. The multiple downcomer fixed valve tray had much larger F_{flooding} than sieve tray. The downward inclined margin on the fixed valve deflected the vapor downward into the liquid. The vapor out of valve side aperture thrust onto the tray deck, did not impinge with vapor from adjacent fixed valves. Additionally, the distance of adjacent fixed valves was larger than V1 valve trays, which also deferred the vapor impingement. These improvements on vapor-liquid contact made the froth steady on tray deck and reduced entrainment. According to Glitsch's design rules[10], F_{flooding} is 2.15 $\text{kg}^{0.5}\cdot\text{m}^{-0.5}\cdot\text{s}^{-1}$ at liquid flow rate of 20 $\text{m}^3\cdot\text{m}^{-2}\cdot\text{h}^{-1}$ and 1.95 $\text{kg}^{0.5}\cdot\text{m}^{-0.5}\cdot\text{s}^{-1}$ at liquid flow rate of 80 $\text{m}^3\cdot\text{m}^{-2}\cdot\text{h}^{-1}$ in a V1 valve tray column with same tray spacing and same relative open area. Comparisons of F_{flooding} between multiple downcomer fixed valve tray and V1 valve tray showed that multiple downcomer fixed valve tray had less entrainment.

3.4. Effect of adding anti-dumping units on impact dumping

To decrease the possibility of the impact dumping from the downcomer into the next downcomer, the anti-dumping units were manufactured and settled at the receiving area on the tray deck. These units were used to guide the liquid from penetrating the orifices and prevent the liquid from dumping into the downcomers of the tray below directly. To test the effect of adding anti-dumping units, experiments were conducted in the 500mm diameter simulator with test tray #4 before and after the anti-dumping units were installed, respectively. Square containers were hung under the receiving area to catch the dumping liquid. The liquid was then piped into the measuring cylinder. Both the experimental results of the impact dumping of the test tray with and without the anti-dumping units are shown in Figure 8. The dumping amount is defined as the flux of the dumping liquid collected under the receiving area based on the cross-sectional area of the column,

$$V_{dump} = \frac{V_{LD}}{\frac{\pi}{4} D^2} \tag{3}$$

The liquid rate was fixed at $51m^3/m^2 \cdot h$ and the F-factor varied from $0.5m/s \cdot (kg/m^3)^{1/2}$ to $2.5m/s \cdot (kg/m^3)^{1/2}$. The solid circles in Figure 8 represent the results of the conventional multiple downcomer tray. When the F-factor was lower than 1.7, the impact dumping amount measured was larger than $4m^3/m^2 \cdot h$, about 8% of the liquid rate. This fraction decreased with the increase of the vapor rate. The impacting dumping amount was measured as about $2m^3/m^2 \cdot h$, 4% of the liquid rate, when the vapor rate increased to $2.5(kg/m^3)^{1/2} \cdot m/s$. As a conclusion, considerable impact dumping can be detected under the tray deck of the conventional multiple downcomer tray. The results of the modified multiple downcomer tray were symbolized by the squares as shown in Figure 8. The largest amount of impact dumping was reduced to $1m^3/m^2 \cdot h$ with the vapor rate at $0.5(kg/m^3)^{1/2} \cdot m/s$. There was almost no impact dumping when vapor rate was a little higher. This is due to the effect that the dumping liquid was partially prevented by the anti-dumping units and partially dropped onto the tray deck below.

3.5. Efficiency comparison between conventional multiple downcomer tray and DJ-3 tray

The so-called DJ-3 tray, improved from conventional multiple downcomer trays with directing baffles, anti-dumping units, and combined packing, was developed. The sketch of its structure was shown in Figure 9. To compare the efficiency between the DJ-3 tray and the conventional multiple downcomer tray, experiments were conducted in the 300mm diameter distillation column with the ethanol-water system at atmospheric pressure under total reflux.

The tray efficiency (E_T) in this paper means the overall tray efficiency, defined as the ratio of the number of theoretical stages (N) over the number of the actual stages (N_T):

$$E_T = \frac{N}{N_T} \tag{4}$$

The results are shown in Figure 10. The F-factor varied from 0.5 to $2.2m/s \cdot (kg/m^3)^{1/2}$. The efficiency of the DJ-3 tray, symbolized by squares, was found to be about 10%-15% higher than that of the test tray #5 without the combined packing and directing baffles. Particularly, when the vapor load was low, a peak in the tray efficiency was observed. The possible reason for the efficiency peak is the elimination of the impact dumping and the presence of mass transfer in the packing. When the vapor F-factor increased to $2\text{-}2.5m/s \cdot (kg/m^3)^{1/2}$, the DJ-3 tray still featured 10%-15% higher efficiency than the test tray without promotions due to the packing layer which prevented the entrainment liquid and provided them extra mass transfer area. The tray efficiency of V-1 valve trays(tray #6), determined in the same experimental condition, is also shown in Figure 8 for comparison. It can also be seen that the DJ-3 tray features higher tray efficiency than the valve trays in this system.

Additional tests were conducted to investigate the effect of adding anti-dumping units on the efficiency of the multiple downcomer tray. The efficiencies of test tray #8 with and without anti-dumping units were determined in the experiments. The results are shown in Figure 11. The F-factor varied from 0.5 to $2.1m/s \cdot (kg/m^3)^{1/2}$. When the vapor load was low, the anti-dumping units showed a large effect on the tray efficiency with an increase of about 20%. The reason for the efficiency peak may be the elimination of the weeping, including impact dumping and random weeping. With the increase of vapor load, the tray efficiency difference between these two trays becomes negligible due to the lower impact dumping at high vapor rates.

3.6. Efficiency comparison between DJ-5 tray and sieve tray, V1 valve tray

The so-called DJ-5 tray, improved from conventional multiple downcomer trays with directing baffles, anti-dumping units, and fixed valve as bubble unit, was developed. The sketch of its structure is shown in Figure 1(f). To compare the efficiency between the DJ-5 tray and the conventional multiple downcomer tray, experiments were conducted in the 600mm diameter distillation column with the cyclohexane/N-Heptane system at atmospheric pressure under total reflux.

The results are shown in Figure 12. The F-factor varied from 0.3 to 2.8m/s·$(kg/m^3)^{1/2}$. The efficiency of DJ-5 tray(tray #9), symbolized by squares, was found to be higher about 20% than that of conventional multiple tray(tray #10) in all vpapr load ranges. One reason is that the entrainment of DJ-5 tray is less than that of conventional multiple tray in the same vapor/liquid load. On DJ-5 tray, the vpour flows horizontally into liquid because of the existing of vlave cover, while on conventional multiple tray, the vapor flows vertically into liquid. Another reason is that the vapor/liquid contacting time on DJ-5 tray is longer than that on conventional multiple tray. On DJ-5 tray, vpour firstly flows horizontally into vapor-liquid mixture, then ascends from the vapour-liquid mixture, while on conventional multiple tray, vapour directly flows vertically through vapour-liquid mixture. When the vapour load is low, the efficiency of DJ-5 tray is less than V1 valve tray (tray #7). The reason is that the free open area is low on V1 vlave tray as some V1 valves don't lift up when the vapour load is low. The froth height on DJ-5 tray is higher than that on conventional multiple tray when the vapour load is low. When the vapour load is larger, all V1 valves lifts up and the free open area of V1 valve tray (tray #7) is same as that of DJ-5 tray (tray #9). The efficiency of DJ-5 tray is as much as V1 valve tray.

4. Conclusions

Several promotion devices including directing baffles, combined packing, fixed valves with downward inclined margin as bubble units and anti-dumping units were used to improve the performance of the multiple downcomer trays. It was shown that uniform liquid distribution was obtained after directing baffles were settled in the simulator. Up to 78% entrainment was captured by the combined packing. Much entrainment was restrained by the fixed valves with downward inclined margin as bubble unit. Almost no dumping liquid was detected when anti-dumping units were used on the tray. It was also observed that the DJ-3 tray equipped with promotion devices featured 10%-15% higher tray efficiency than that of the conventional multiple downcomer trays in the ethanol/water system. The experiment with cyclohexane/N-heptane system showed that DJ-5 tray has characteristics of 10%~ 20% higher tray efficiency than that of the conventional multiple tray in all operating load and same tray efficiency with V1 valve tray in middle and large operating load.

Nomenclature

D	Column diameter, m
E_T	Overall tray (column) efficiency, %
e_v	Entrainment, kg water/kg air
F	Capacity factor based on cross sectional area, $(m/s)(kg/m^3)^{1/2}$
M_{Le}	Mass metric rate of liquid entrained by ascending gas, kg/h
M_G	Mass metric rate of gas, kg/h
N	Theoretical stages of column
N_T	Actual stages of column
V_{dump}	Dumping amount, m^3/m^2h
V_G	Volume metric rate of gas, m^3/s
V_{LD}	Volume metric rate of dumping liquid under receiving area, m^3/h

Greek symbol

ρ_G Gas density, kg/m^3

Literature Cited

(1). Delnicki, W. V.; Wagner, J. L. Performance of multiple downcomer trays. *Chem. Eng. Prog*. **1970**. 66, 50.

(2). Summers, D. R.; Coleman, S. T.; Venner, R. M. Ethylene fractionator revamp results in 25% capacity increase. *Oil Gas J.* **1992**. August 10, 52

(3). Resetarits, M.R.; Miller, R. J. Increase demethanizer efficiency retray and surrogate rings increase feed rate and capacity. *Hydrocarb. Process.* **1991**. October, 79.

(4). Yu, X.; Xu, C.; Zhu J.; et al. Rectangular suspending downcomer directing tray. US6299146B1. October 9, 2001

(5). Wei K. Study and structure improvement of the multiple downcomer tray. Master Thesis, Zhejiang University of Technology, China, **1998**. (in Chinese)

(6). Wang, W. Study of the performance of the DJ-3 tray. Master Thesis, Zhejiang University of Technology, China, **1988**. (in Chinese)

(7). Yu, X.; Tian J.; Zheng Z.; et al. Combined packing-tray in a vapor-liquid contacting tower and a process for effecting operation with high capacity and high turndown ratio. US6345811B1. February 12, 2002

(9). Yao, K.; Wang, L.; Yu X. Study of the dumping on the tray with hanging downcomer. *Journal of Chemical Engineering of Chinese Universities*. **2002**. 16, 135. (in Chinese)

(10). Shi Jun, Wang Jia-ding, Yu Guo-cong, Chen Min-heng, Handbook of chemical engineering, Chemical Industry Press, Beijing(1996). (in Chinese)

Table 1 Test tray dimensions

Test tray No.	1	2	3	4	5	6	7	8	9	10
Tray diameter (mm)	1200	300	600	500	300	300	600	300	600	600
Number of DC*	1	1	1	1	1	1	1	1	1	1
Length of DC (mm)	1100	270	550	450	270	205	410	270	550	550
Width of DC (mm)	100	50	50	80	50	/	/	50	50	50
Depth of DC (mm)	220	200	400	400	200	260	400	200	400	400
Open area (%)	5.90	16.48	8.97	9.11	16.48	9.11	9.42	16.48	8.97	9.07
Hole diameter (mm)	10	6	/	8	6	39	39	6	/	8
Fixed valve area (mm^2)	/	/	338	/	/	/	/	/	338	/

*: DC is the abbreviation of downcomer.

LIST OF FIGURES:

Figure 1. Sketch of test trays in the experiments.
Figure 2. Schematic diagram of simulator to test entrainment and dumping.
Figure 3. Sketch of fixed valves with downward inclined margins on tray deck.
Figure 4. Schematic diagram of distillation column to test tray efficiency.
Figure 5. The effect of adding directing baffle on distribution of equal mean residence time (in seconds) on a quarter of the test tray #1.
Figure 6. Effect of adding combined packing on entrainment of the test tray #2.
Figure 7. Effect of liquid and gas on entrainment of multiple downcomer fixed valve tray.
Figure 8. Effect of adding anti-dumping units on dumping amount of the test tray #3.
Figure 9. Sketch of DJ-3 tray.
Figure 10. Comparison of overall tray efficiency between test tray #4 with and without combined packing and directing baffles, and test tray #5.
Figure 11. Effect of adding anti-dumping units on overall tray efficiency of the test tray #6.
Figure 12. Comparison of overall tray efficiency between DJ-5 tray and conventional multiple tray, V1 valve tray

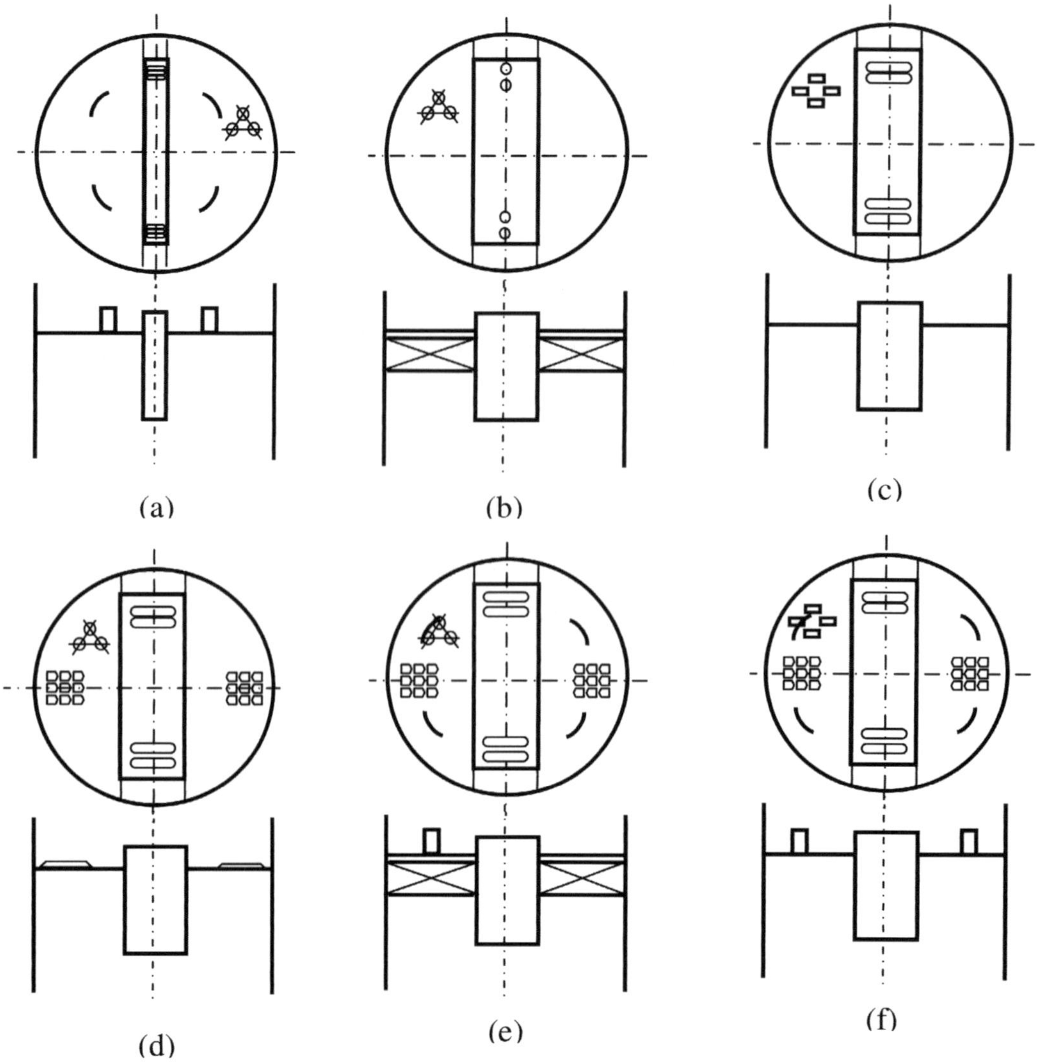

Figure 1. Sketch of test trays in the experiments.

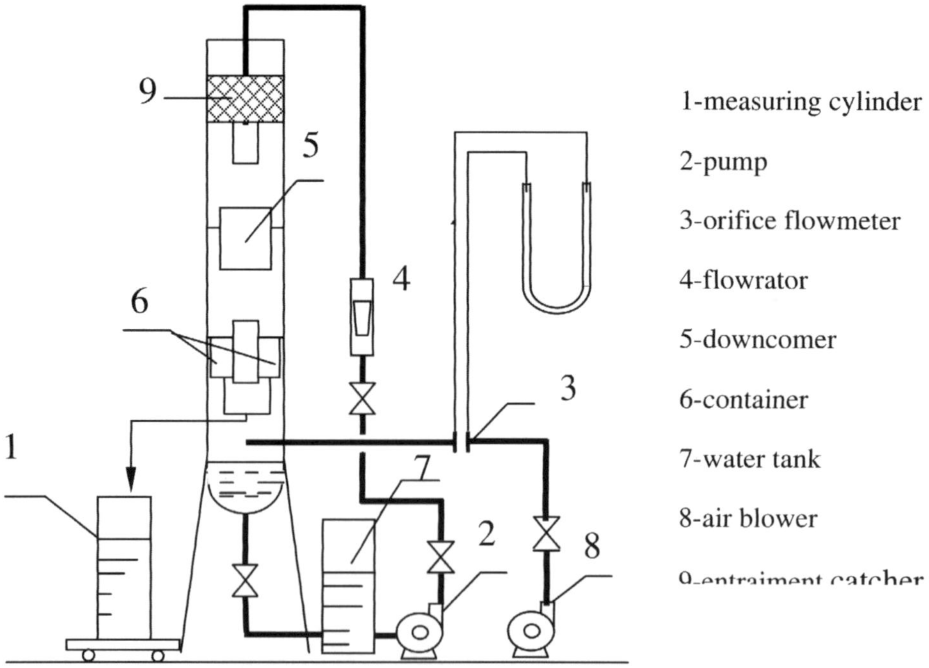

Figure 2. Schematic diagram of simulator to test entrainment and dumping.

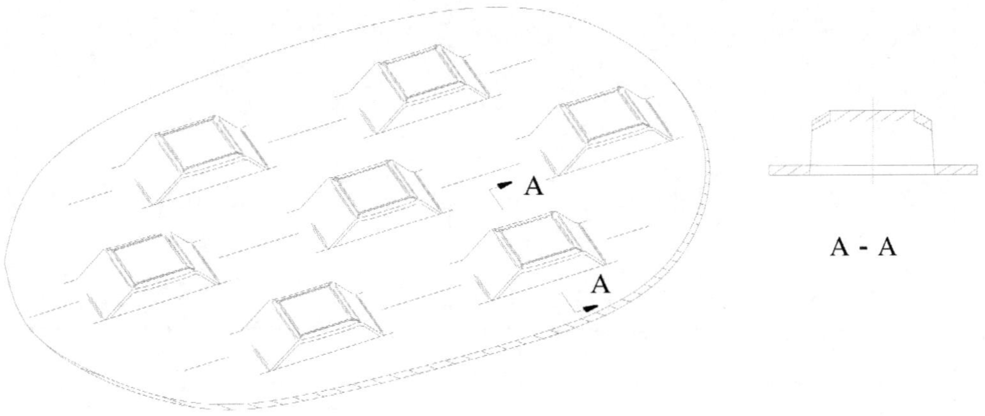

Figure3 Sketch of fixed valves with downward inclined margins on tray deck

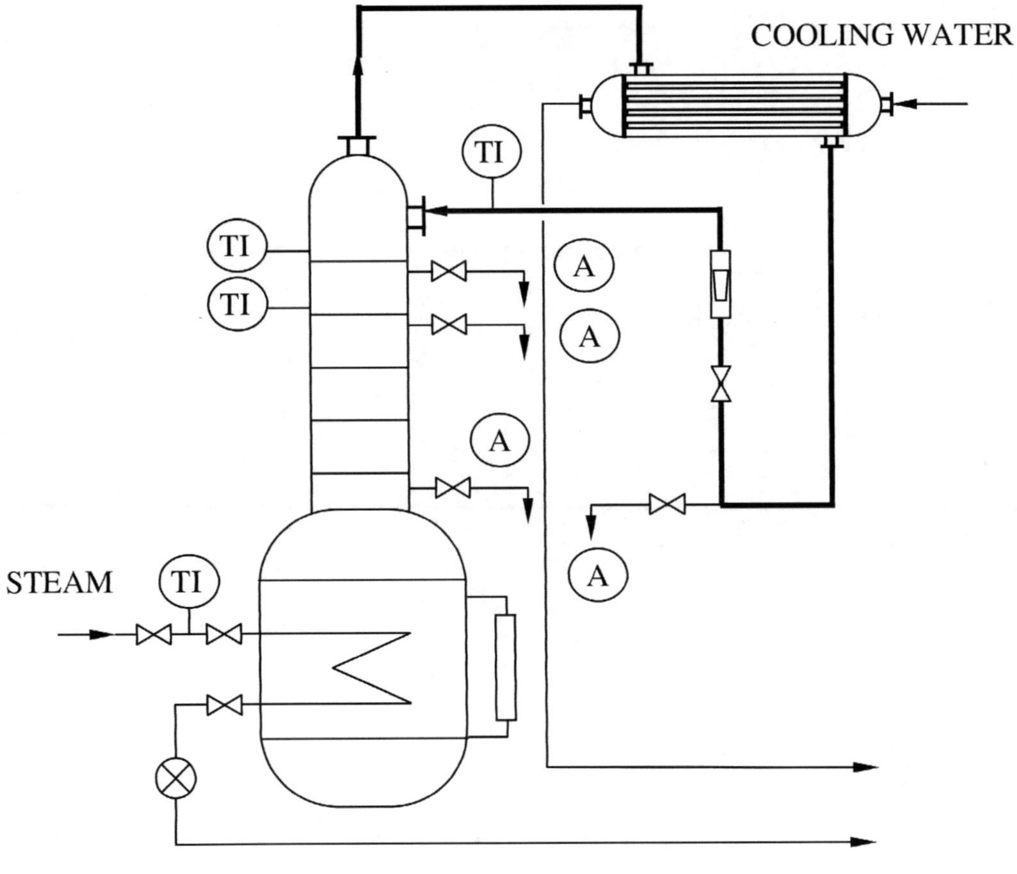

Figure 4. Schematic diagram of distillation column to test tray efficiency.

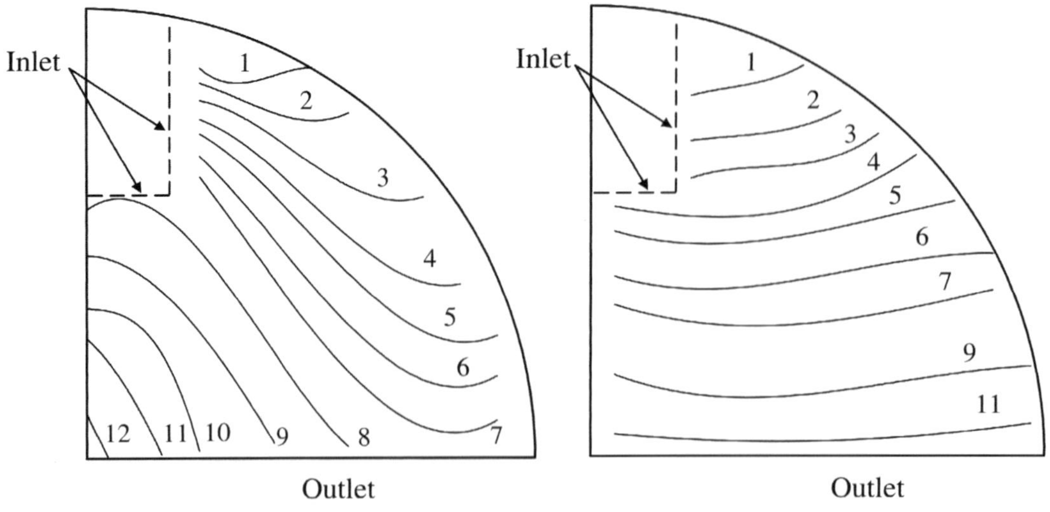

(a) without directing baffles (b) with directing baffles

Figure 5. The effect of adding directing baffle on distribution of equal mean residence time (in seconds) on a quarter of the test tray #1.

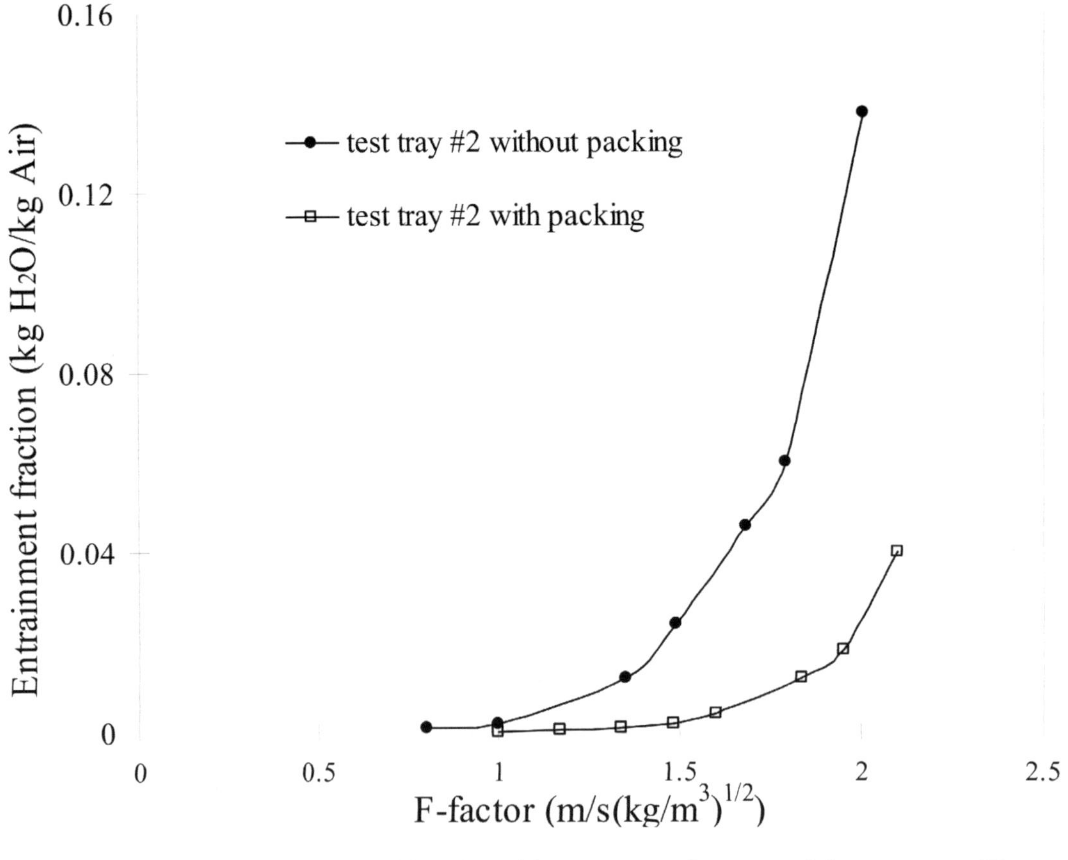

Figure 6. Effect of adding combined packing on entrainment of the test tray #2.

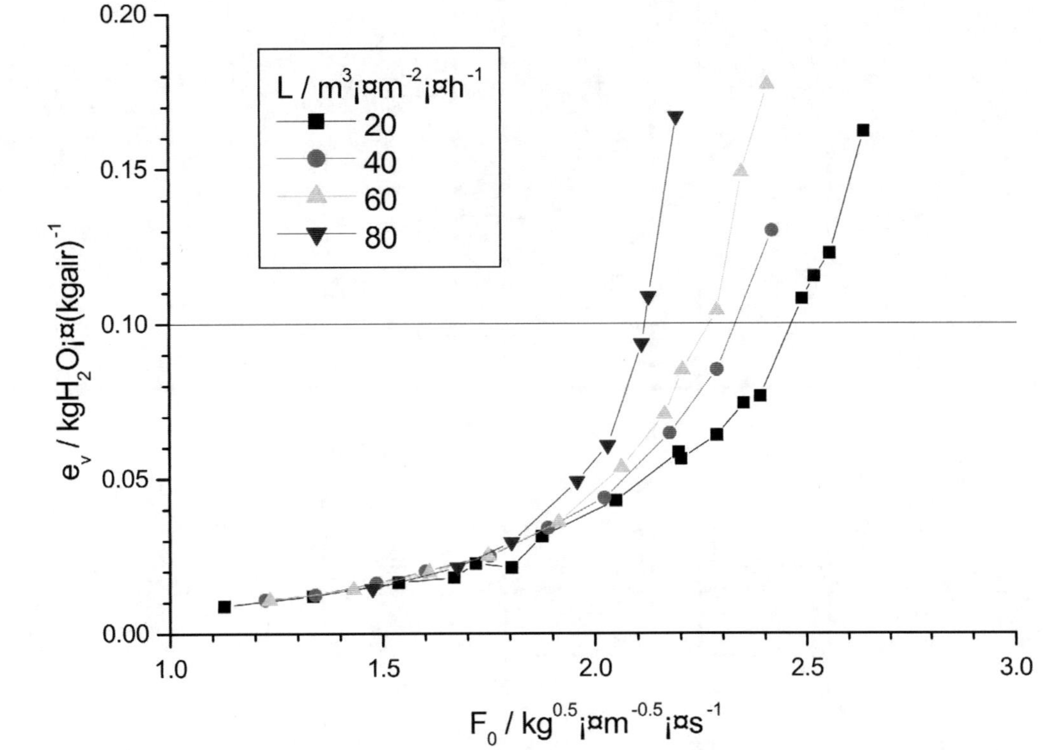

Fig.7 Effect of liquid and gas on entrainment of multiple downcomer fixed valve tray

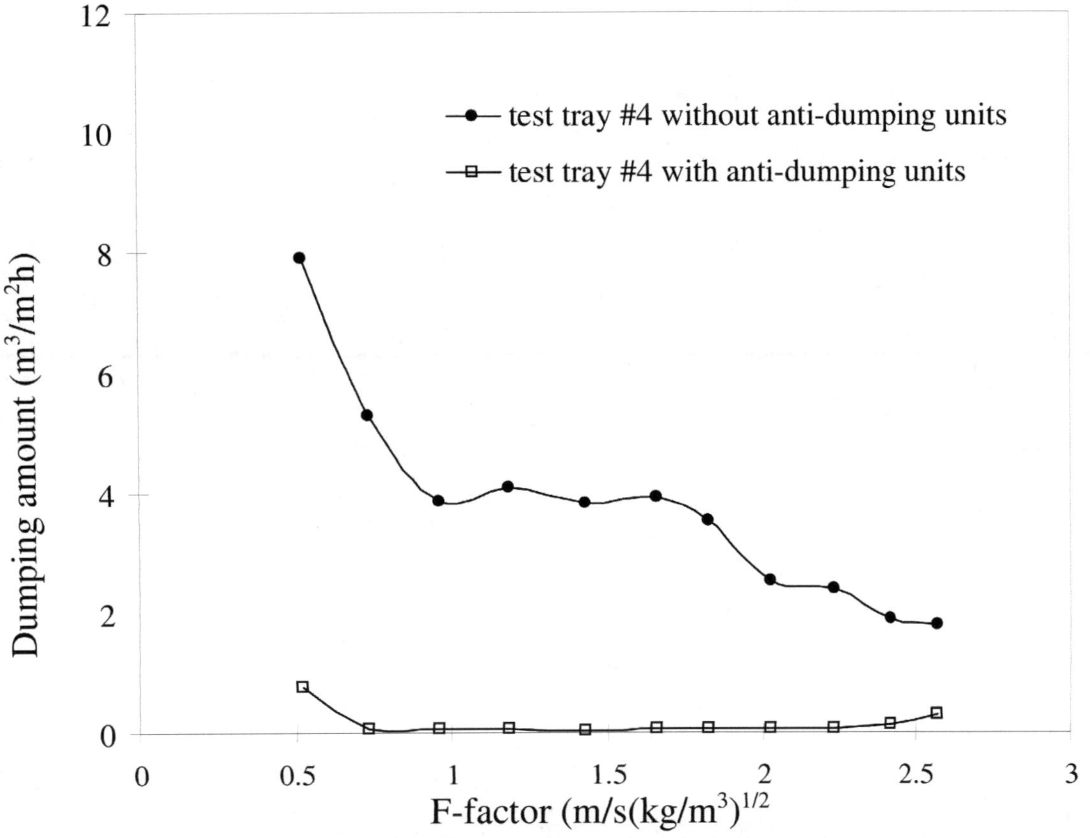

Figure 8. Effect of adding anti-dumping units on dumping amount of the test tray #4.

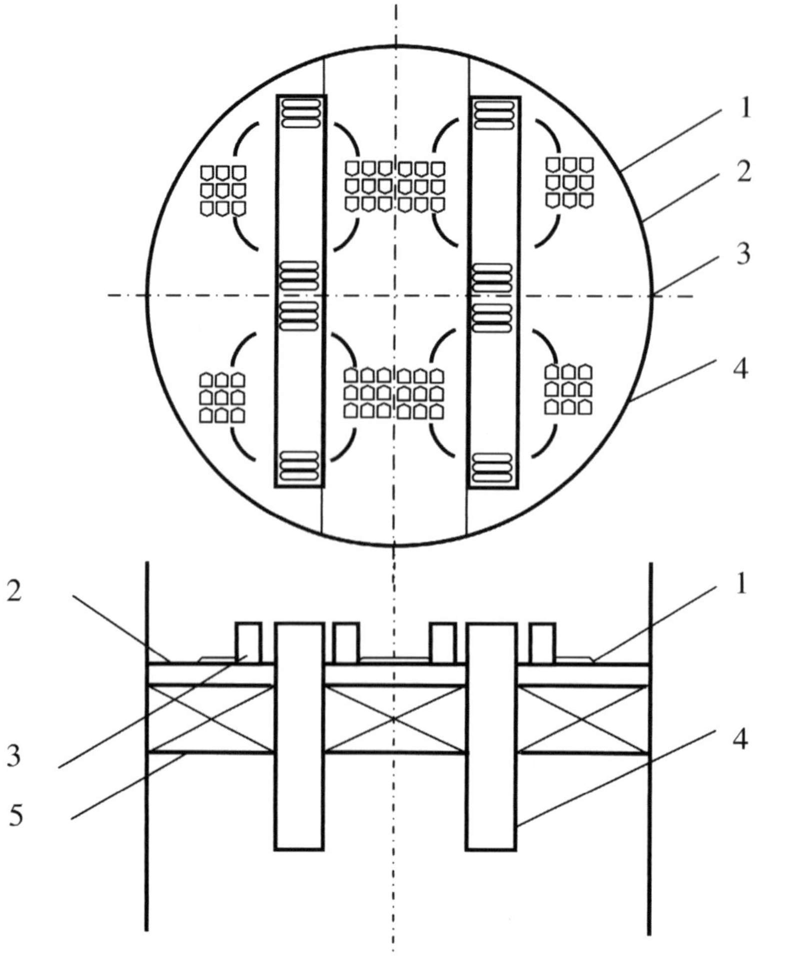

1 anti-dumping units; 2 tray deck; 3 directing baffles; 4 downcomers; 5 combind packing

Figure 9. Sketch of DJ-3 tray

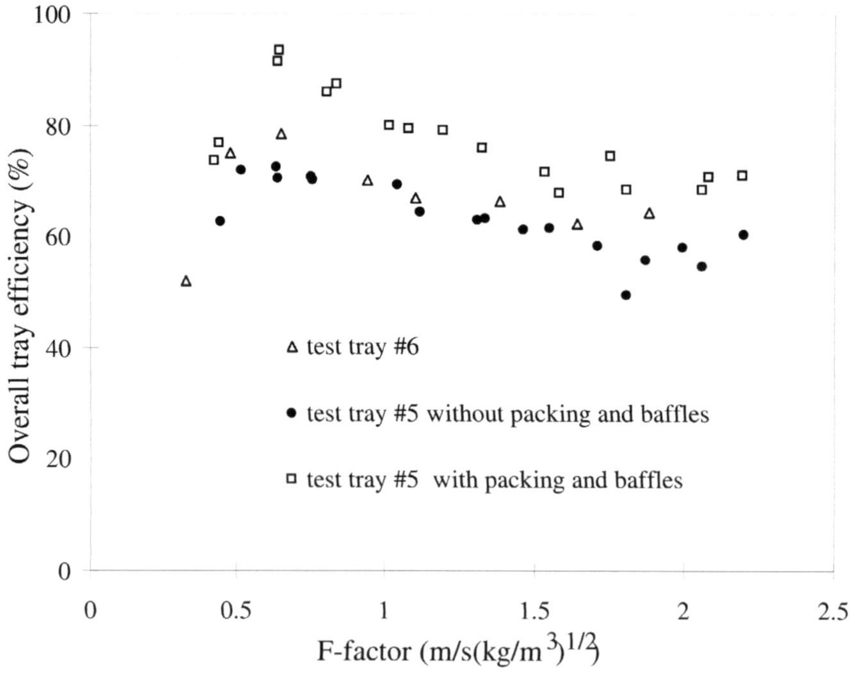

Figure 10. Comparison of overall tray efficiency between test tray #5 with and without combined packing and directing baffles, and test tray #6.

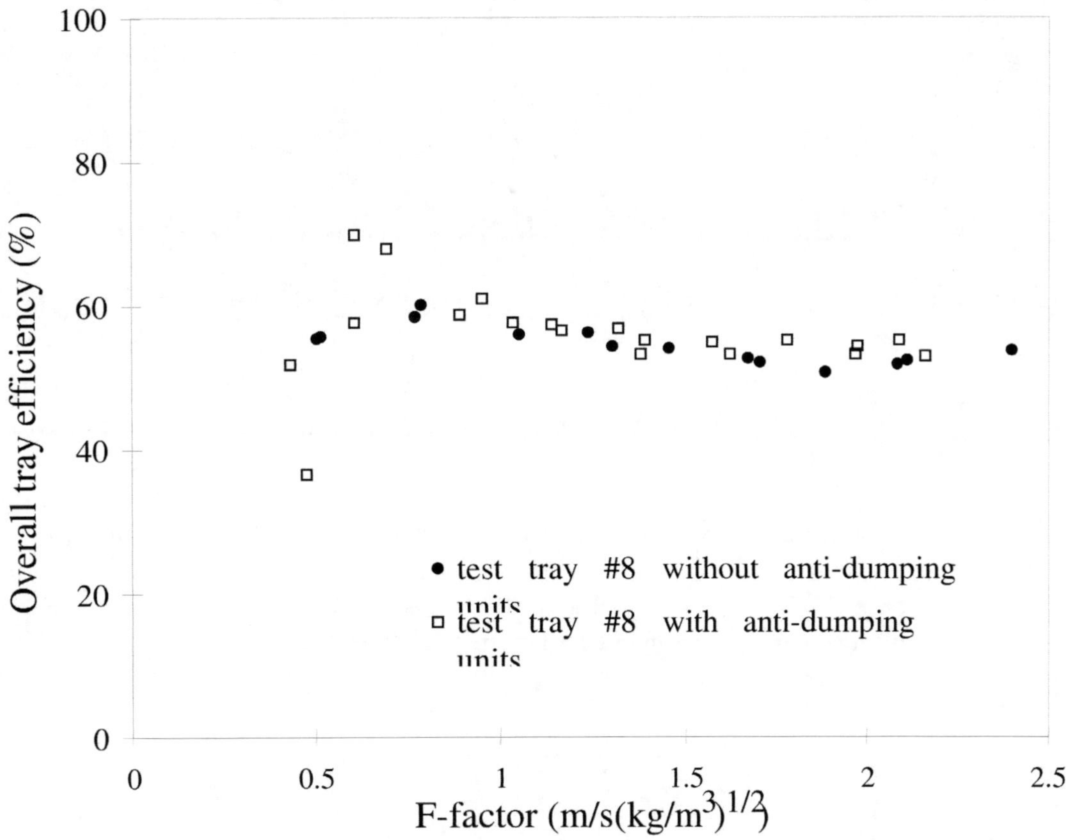

Figure 11. Effect of adding anti-dumping units on overall tray efficiency of the test tray

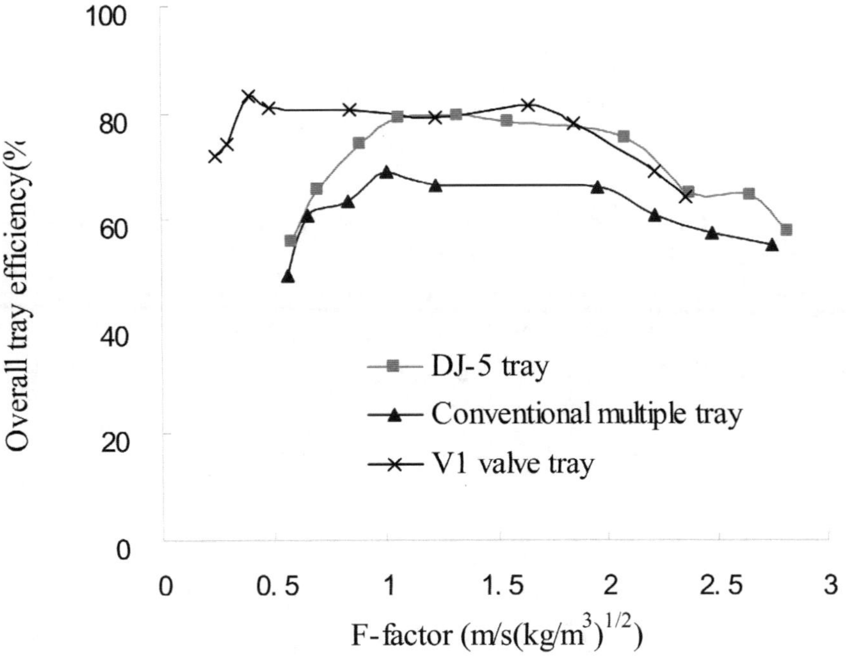

Fig.12 Comparison of overall tray efficiency between DJ-5 tray and conventional multiple tray, V1 valve tray

Study on Two-phase Flow in Tray Downcomers – Backmixing and Choking tests

Xiaomei Yu,Lingyu Zhu, Kejian Yao*

State Key Laboratory Breeding Base of Green Chemistry Synthesis Technology

College of Chemical Engineering & Materials Science

Zhejiang University of Technology

Hangzhou, 310032

P. R. China

---

* Corresponding author. Tel: 86-571-88036778; Fax: 86-571-88033331; Email: yuxm@zjut.edu.cn.

Study on Two-phase Flow in Tray Downcomer - Backmixing and Choking Tests

YU Xiaomei *, ZHU Lingyu, YAO Kejian

College of Chemical Engineering and Material Science , Zhejiang University of Technology , Hangzhou, 310014, China

Abstract To decide the upper limit of tray downcomer, characteristics of V-L two-phase flow in the industrial scale downcomer are studied in the paper. The liquid backmixing and circular flow in downcomer are tested with the fluoro-tracer agent and the hot water tracer technique in a 1900mm and a 1200mm diameter simulator , respectively. The critical chocking liquid-loading intensity of downcomer is determined in a rectangle simulator, of which the length is 800mm and the width is 600mm. The results shows severe liquid backmixing in the downcomer with backmixing number are higher than 0.1 and the dimensionless variance are about 0.5. The liquid circular flow is observed in the middle of downcomer. It is also shown that the value of critical chocking liquid-loading intensity of downcomer is less than the value predicted by previous empirical correlation. The results obtained in the large-scale experimental apparatus should be more reliable than that in small-scale equipment. It can provide basic data for engineering design.

Keywords downcomer, backmixing, circular flow, critical liquid-loading intensity, chocking

1. INTRODUCTION[1]

Tray column have wide applications in industry which occupies absolute predominance in pressure distillations and absorptions, and in column with multi-input/output, and for block up material[1]. Downcomer is an important part which enables liquid to descend fluently from one tray onto the next tray below and provide a space for bubbles to segregate from the liquid overflowed weir. Adequate downcomer area and weir length must be provided to ensure liquid capacity and bubble space for mass transfer. Therefore, it is a key step for designers to decide downcomer size reasonably in accordance with the characteristic of two-phase flow in downcomer[2,3].

The downcomer flooding and the jet flooding are two major factors limited the hydraulic performance of trays. For trays operated at pressure or at high liquid-vapor ratio, which is commonly known in emulsion regime, the downcomer flooding is always the upper limit of tray performance.

S. C. Kler *et al.* [4] inspected the performance of the tray in a hydrocarbon condensate stripper of olefins plant operated at high pressure. They found that the vapor-liquid mixture in the downcomer behaves like a foaming system so that for high pressure systems in the emulsion flow regime and for foaming systems, the assumption of no phase separation in the downcomer is a satisfactory basis for determining tray capacity. For systems in pressure operation or foaming, the gas hold-up is higher than others. It showed that the gas is carried down seriously by liquid, as limited the liquid capacity of downcomer obviously. Yu studied on the gas hold-up of the downcomer in emulsion regime[5], and verified the results of S. C. Kler, *et al.* and FRI.

All of results above are based on the assumption of gas backmixing and liquid piston flow in downcomer, so did design criteria to determine the upper limit of downcomer. In recent years, some researchers had reported their studies on non-piston flow feature in downcomer[6]. Wang *et al.*[7] investigated the liquid mixing in the segment downcomer and suggested that the liquid is partially mixing and the mixing varied on the weir length. This paper presents an exploration of the liquid backmixing, circular flow and chocking of downcomer in the industrial scale columns. The results provide some rule for designing of large downcomer.

2. EXPERIMENTAL APPARATUS

The experiments of backmixing in downcomer are carried out in a 1900 mm diameter simulator with air/water system. The residence time distribution of the fluid is tested by the fibre optics technique. Schematic diagram of the experimental apparatus is illustrated in Figure 1. The sketch of the test DJ tray used in the simulator, which is equipped with two rectangular downcomers, is shown in Figure 2. Corresponding to the three liquid spouts in downcomer, five fibre-optics probes are settled on the test tray.

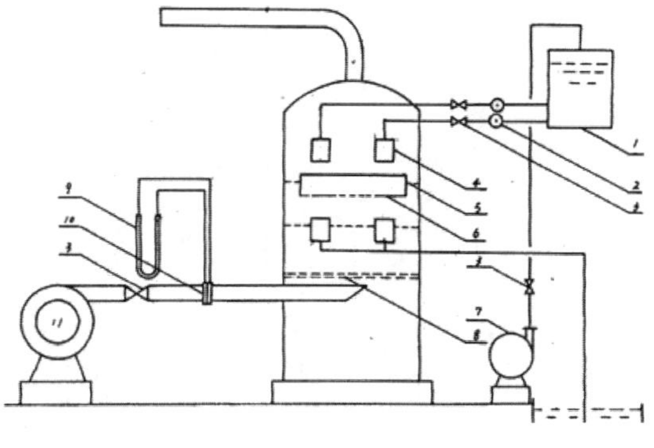

* To whom correspondence should be addressed. E-mail: yuxm@septech.com.cn

1.water tank 2.vortex- meter 3.valve 4.liquid distributor 5.test tray 6. downcomer 7. water pump 8. gas distributed plate 9.manometer 10.orifice meter

Figure 1. Schematic diagram of experimental apparatus.

The liquid backmixing is also tested in a 1200mm diameter simulator with a DJ tray. Hot water tracer is used to determine the distribution of liquid resident time. The schematic diagram of the experiment diagram is illustrated in Figure 3.

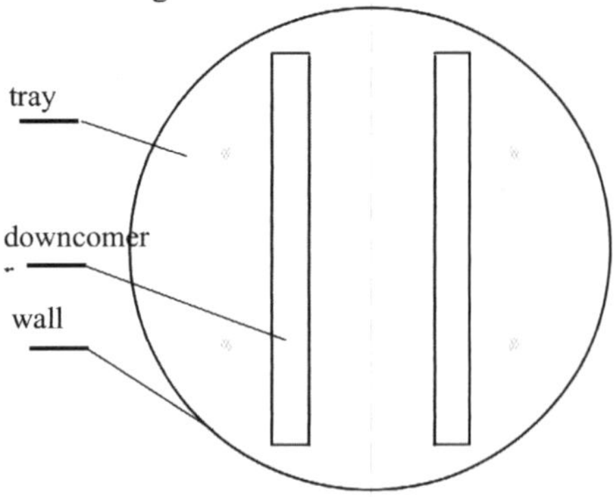

Figure 2. Sketch of test DJ tray.

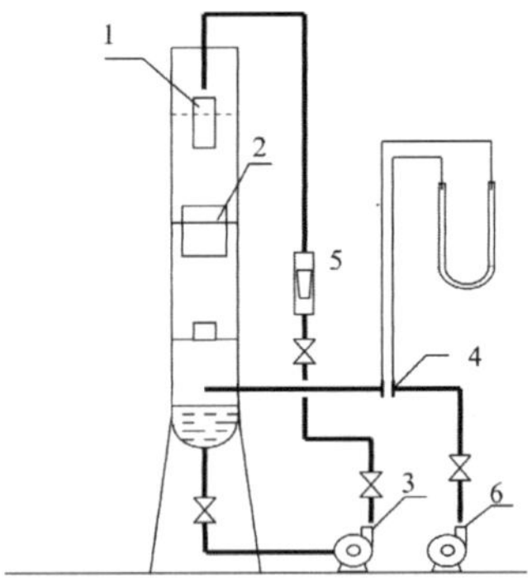

1. liquid distributor 2. test tray 3 - pump ; 4 - orifice flowmeter and differential manometer 5 - flowrator 6 - air blower

Figure 3. Schematic diagram of experimental apparatus.

The experiments of chocking are conducted in a series of rectangular downcomers. The experiment apparatus arrangement is shown in Figure 4. The widths of downcomers are 80mm and weir length are from 300mm to 600mm, respectively.

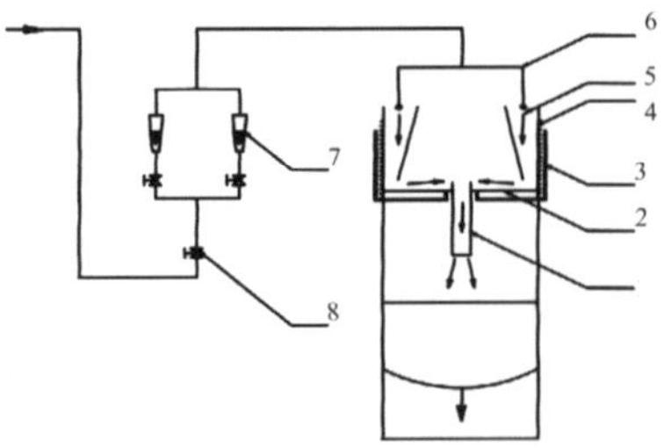

1. downcomer 2. tested tray 3. manometer 4. measuring tape 5. liquid-directing tube 6. liquid-distributor 7.rotameter 8. valve

Figure 4. Schematic diagram of the experimental apparatus.

3. RESULTS AND DISCUSSIONS

3.1 Backmixing and circular flow in downcomer

The backmixing and the circular flow in downcomer are significant to upper limit of the downcomer capacity. Figure 5 showed that the flow regime on the tray is typical foaming.

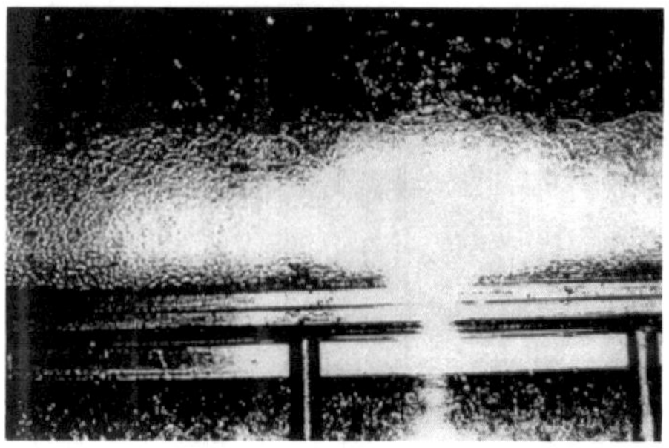

Figure 5. Foam on 1900mm diameter tray.

Liquid backmixing in the downcomer is also tested. First experiment to determine the backmixing and the circular flow in downcomer is carried out in the 1900mm diameter column. The RTD curve obtained in the middle of the downcomer in Figure 6 is characterized by long tail and two peaks, which indicated the existence of circular flow.

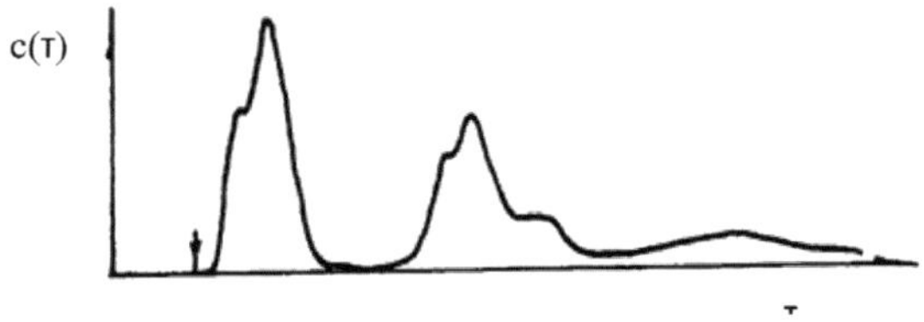

Figure 6. RTD curve in the downcomer of test tray.

Using the general moment method, t^r is taken as the weighted factor. This factor magnifies the tail error and affects the second-order moment intensively. However, using the transfer function method with e^{-st} as the weighted factor, it will eliminate tail error and magnify the head error. It is decided that a modified moment method is to be used here for data treatment.

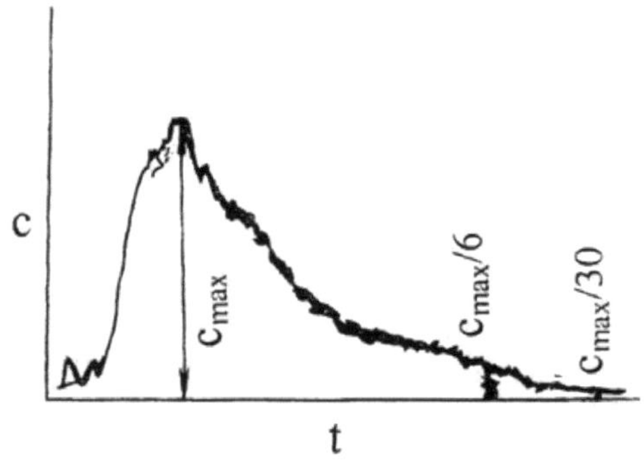

Figure 7. Data treatment.

As shown in Figure 7 that The RTD curves could be split into two sections: an initial section which could be numerically analyzed, and a long tail section which could be closely described by the following equation

$$c = c_A e^{-k(t-t_A)} \quad (1)$$

where (c_A, t_A) is the coordination of a point on the curve and k is a decay factory. This second section of the curve ranges between the values of t corresponding to the values of c equal to $c_{max}/30$ and $c_{max}/6$, c_{max} being the peak value.

The first portion of the curve is evaluated by splitting it into eight subsections. Taking interval Δt as one single step, numerical integration can be carried out using the Simpson method. For the tail section Eq.(1) is used for integration. Then, the following equations are obtained

$$A = \sum_{i=1}^{8} \frac{c_i + 4c_{i+1}}{3} \cdot \frac{t_1 - t_0}{8} \quad (2)$$

$$\bar{t} = \frac{1}{A}\left[\sum_{i=1}^{8} \frac{c_i t_i (t_1 - t_0)}{8} + \frac{c_1 - c_2}{k}(t_c + \frac{1}{k})\right] \quad (3)$$

$$\sigma_t^2 = \frac{1}{A}\left[\sum_{i=1}^{8} \frac{c_i t_i^2 (t_1 - t_0)}{8} + cc\right] - \bar{t}^2 \quad (4)$$

$$\Pr = \frac{2}{\sigma_t^2} \quad (5)$$

Where σ_t^2 is the relative time square deviation, Pe is the Peclet Number, $Pe = uL / D_E$, t_0 is the time at the starting point, t_1 is the time corresponding to $c_1 = c_{max}/6$, t_2 is the time corresponding to $c_2 = c_{max}/30$. Other parameters in formula are as follows.

$$t_c = \frac{c_1 t_1 - c_2 t_2}{c_1 - c_2} \quad (6)$$

$$k = \frac{\ln(c_1 - c_2)}{(t_2 - t_0)/(t_1 - t_0)} \quad (7)$$

$$cc = \frac{c_1 - c_2}{k}(t_c^2 + \frac{2t_c}{k} + \frac{2}{k^2}) \quad (8)$$

According to the above equations, the experimental data at five measuring points under four different operations are correlated.

The remaining operating conditions are listed in Table 1.The results listed in Table 1 shows that the backmixing number are higher than 0.1, about 0.1323~0.1596 in four runs. The backmixing number is the reciprocal of Peclet number. which shows the level of backmixing. Therefore, it is shown that there exists severe liquid backmixing in the entire downcomer, and the more liquid rate the more liquid backmixing.

The liquid backmixing in the downcomer is also tested in the 1200mm diameter simulator with hot-water tracer. The operating conditions and the results of the experiments are listed in Table.2. It shows the same order of magnitude of liquid backmixing in the large-scale downcomer as on the tray deck. The dimensionless variance (σ^2/τ^2) are about 0.5~0.6. It can on longer to be ignored.

3.2 Chocking in downcomer

Besides the downcomer backup flooding and the downcomer velocity flooding, liquid blocking up at the downcomer entrance (Chocking) can also result in the breakdown of downcomer. Downcomer choking is studied in both single liquid phase and two-phase condition. The critical chocking liquid-loading intensity is determined in the rectangle simulator with air/water system. Table 3 shows the value of critical chocking liquid-loading intensity is less than that predicted by previous correlation in small-scale downcomer. The reason for the premature chocking should be the existence of circular flow and backmixing of liquid in the large-scale downcomers.

The critical liquid velocity of chocking, which can be calculated by critical liquid-loading intensity, is about 0.3m/s. Figure 8 showed the phenomena of liquid loading the weir from overflow to chocking.

Figure 9 compares the correlated critical liquid velocity with the experimental value. It is much higher than the upper limit of velocity (0.2m/s) in downcomer commonly used in commercial tray design.

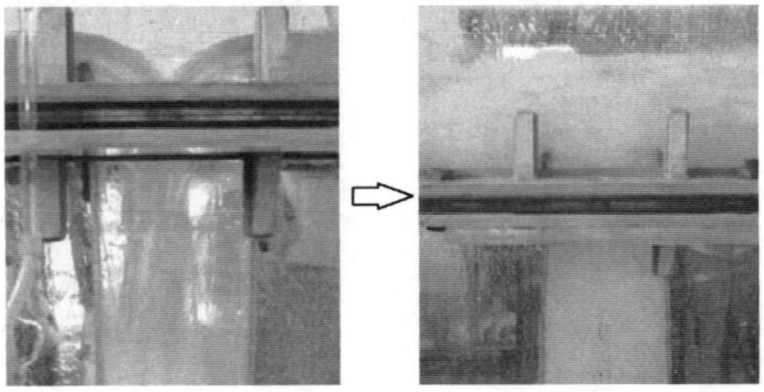

Figure 8. Liquid-loading on the weir.

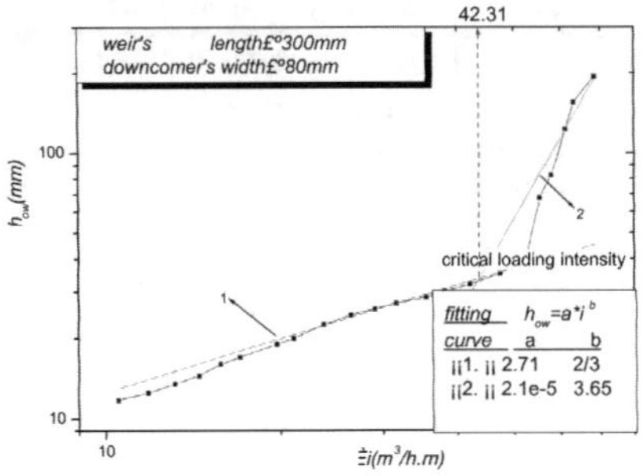

Figure 9. Formula of h_{ow} & i.

Table 1. Experimental results on the φ1900 column

Rate of air $[m^3h^{-1}]$	Rate of water $[m^3h^{-1}]$	Variance σ^2	Peclet Number Pe	Backmixing Number
10000	20	0.4048	7.556	0.1323
12000	20	0.4840	6.627	0.1509
10000	30	0.5025	6.449	0.1551
12000	30	0.5228	6.267	0.1596

Table 2. Experimental results on the φ1200 column

Rate of air $[m^3 h^{-1}]$	Rate of water $[m^3h^{-1}]$	Flow parameter (FP)	σ^2/τ^2
2788	20	0.2343	0.4927
3720	20	0.2928	0.4918
2788	25	0.1757	0.5049
3720	25	0.2196	0.4890

Table 3. Compared results of critical liquid loading intensity

Weir length/mm×Downcomer width / mm	600×80	550×80	500×80	450×80	400×80	350×80	300×80
i_c / $m^3\cdot(h\cdot m)^{-1}$ by this paper $i_c = 0.081(\frac{n}{n+1})^{1.5}b^{1.5}$	45.73	45.32	45.44	44.90	43.88	42.69	42.31
I_c / $m^3\cdot(h\cdot m)^{-1}$ by previous $I_c = 0.101(\frac{n}{n+1})^{1.5}b^{1.5}$	59.90	58.95	57.84	56.54	54.98	53.07	50.69
i_c / I_c	0.763	0.769	0.786	0.794	0.798	0.804	0.835
u_{Lmax} / ms^{-1}	0.318	0.315	0.316	0.312	0.305	0.296	0.294

4 CONCLUSIONS

(1) In the large-scale downcomer exists obvious liquid circular flow and severe liquid backmixing. The number of backmixing are about 0.1323~0.1596. The dimensionless variance is about 0.5.

(2) The values of chocking critical liquid-loading intensity determined in the downcomer of 80mm width and 300~600mm weir length are less than that obtained previously in the small-scale downcomer. The results show a premature chocking.

(3) The critical liquid velocity of chocking, which can be calculated by critical liquid-loading intensity, is about 0.3m/s. it is much higher than the upper limit of velocity (0.2m/s) in downcomer commonly used in commercial tray design.

REFERENCES

1 JI Jianbing, Study on Performance of fluid Dynamics and Mass Transfer of Compound Tray, Doctor Paper, 2001

2 XU Chongsi, The effect of downcomer on tray characteristics, Petro-Chemical Equipment[J],1982,5,P57-62

3 XU Chongsi, YU Xiaomei, Rational design for width of Rectangular downcomer, Petro-Chemical Equipment [J] Petro-Chemical Equipment[J],1980,1,P25-29

4 S. C. Kler , R. J. P. Brierley and M. C. G. Del Cerro, I. Chem. E. Symposium Series No.104 B391-399

5 YU Xiaomei, WU Renan, XU Chongsi, Gas Hold-up Distribution in Segment Downcomer, J. of Chem. Eng. of Chinese Universities[J], 9(4),1995,p332-337

6 YU Xiaomei, TIAN Jun, Liquid backmixing and circulation in rectangular downcomer, Chinese J. of Chem. Eng. , 5(4) 371-375 (1997)

7 Wang Zhongcheng, et al, Chem. Eng. 22(2),1994(Chinese)

The Performance of Random Packing for High Pressure Distillation Systems

John Kunesh[1], Frits Zuiderweg[2], Tony Cai[2], A. Shariat[2], D.W. King[2], and **G.X. Chen**[2]. (1) Past Chair, Separations Division, Consultant, P.O. Box 377, Red River, NM 87558, (2) Fractionation Research, Inc., P.O. Box 2108, Stillwater, OK 74076

The experimental results of 50 mm Pall Rings with high pressure systems are presented and discussed. The packing was tested in a 1.22 m high pressure distillation column with a bed length of 3.66 m at Fractionation Research, Inc. (FRI). The tests were conducted with the iC4/nC4 system at various column pressures of 11.4 to 34.4 bar. The experimental results show unusual HETP trend and higher HETP than expected. The measured data also have a high degree of scatter for high pressure systems. The data are compared with those of the Nutter Ring No. 2.5 with the same system and operational pressures. It is found that the distributors affect the packing performance significantly. The possible causes of the unexpected HETP and unusual trend are discussed including the effect of samplers, vapor back-mixing, and condensation. Finally, a larger than usual safety factor is suggested for the random packing application at high pressure systems.

Reaching New Performance Levels with Surface Enhanced Raschig Super-Pak Structured Packings

Michael Schultes, Raschig Ring Department, RASCHIG GMBH, Mundenheimerstr. 100, Ludwigshafen, Germany and **Simon Chambers**, Raschig-Ringe, RASCHIG GMBH, Mundenheimerstr. 100, Ludwigshafen, Germany.

In order to penetrate the popular 250 m2/m3 structured packing market, a technically superior and commercially viable alternative to Raschig Super-Pak-300 (RSP-300) was developed: the Raschig Super-Pak-250 (RSP-250 wSE). It has a surface area of 250 m2/m3. The rows of sinusoidal waves within vertical packing sheets are surface enhanced to encourage greater turbulent radial spread of thin liquid film flows on the front and back of the waves on each sheet within an element. RSP-250 wSE was tested at SRP, University of Texas at Austin under identical total reflux conditions to RSP-300 [1] on the cyclohexane/n-hepane system at four operating pressures to characterise performance.

The effect of surface enhancement on the mass transfer efficiency of RSP-250 wSE compared to Raschig Super-Pak 300 without surface enhancement (RSP-300 woSE) is significant with consistently lower HETP values. On comparing with the Montz B1-250 and B1-250M [2] as well as F.R.I. tested Sulzer Mellapak M250Y and MellapakPlus M252Y at 0.34 and 1.65 bar (5 and 24 psia) system pressures [3], the hydraulic-efficiency RSP-250 wSE results are outstanding.

For all system pressures, the pressure drop of RSP-250 wSE are consistently lower by an order of magnitude than the 250 m2/m3 surface area standard structured packings and significantly less than the high capacity types at high flow rates. Combined with excellent efficiencies, the pressure drop per theoretical stage is even more impressive.

Given the outstanding hydraulic and mass transfer efficiency performance, the credentials of RSP-250 wSE competing in the 250 m2/m3 surface area structured packing market can now be firmly established.

References [1]Schultes, M., Chambers, S., "How To Surpass Conventional and High Capacity Structured Packings with Raschig Super-Pak", AIChE Spring National Meeting, Orlando, FL, 23-27 April, 2006. [2]Olujic, Z., Seibert, A.F., Kaibel, B., Jansen, H., Rietfort, T., Zich. E., "Performance of a New High Capacity Structured Packing", AIChE Spring National Meeting, Houston, TX,22-26 April, 2001. [3]Pilling, M., Spiegel, L., "Design Characteristics and Test Validation of High Performance Structured Packing", Distillation Symposium, Distillation Honors John Kunesh, AIChE November National Meeting, Reno Nevada, November 4-9, 2001.

Experimental Characterization of High Performance Montz Structured Packings

Zarko Olujic[1], Thomas Rietfort[2], Helmut Jansen[2], Bjoern Kaibel[2], Egon Zich[2], Gerhard Ruffert[3], and T. Zielke[3]. (1) Delft University of Technology, Leeghwaterstraat 44, Delft, 2628 CA, The Netherlands, (2) Engineering, Julius Montz GmbH, Hofstrasse 82, Hilden, 40723, Germany, (3) Bayer Technology Services GmbH, Leverkusen, Germany

Upon completing screening tests using total reflux facilities available at TU Delft, structured packing geometry refinements have been carried out and representative sizes of new Montz gauze (A3), extended metal (BS), and sheet metal (B1) series packings exposed to thorough testing using state of the art new facility available at Bayer Technology Services. The heart of this installation is a 0.6 m ID, total reflux column, which allows packed heights up to 4 m, and is operated with chlorobenzene/ethylbenzene system at 100 and/or 1000 mbar. A detailed description of relevant features of this installation and the adopted experimental procedure is given. Obtained results, i.e. corresponding hydraulic and mass transfer performances indicate that by combining appropriately the dimensions and design of corrugations and the corrugation inclination angle, i.e. by affecting the pressure drop accordingly, either efficiency or capacity, or even both can be enhanced significantly, depending on the type of the packing.

Experiences with the Quality Rating and Evaluation of Liquid Distributors

Jefferey Scherffius and Henry Z. Kister. Fluor Corporation, 3 Polaris Way, Aliso Viejo, CA 92698

This paper originates describes experiences with using circle diagrams and the Moore and Rukovena Distribution Quality Index method for evaluating liquid distributors. This method was found invaluable in evaluating and improving liquid distributors, but sometimes its application raises interesting questions which will be addressed.

Performance of Liquids Recovery Plant Demethanizer

Glenn Shiveler, Applications - Gas Processing Group, Sulzer Chemtech USA, Inc., 4019 South Jackson Ave., Tulsa, OK 74107 and Derrick Oneal, Thomas Russell Co., 7050 South Yale Avenue, Suite 210, Tulsa, OK 74136.

The performance of a gas plant with a cryogenic Demethanizer equipped with several sections of mondern random packing and tower internals is examined to show the mass transfer separation performance, ethane recovery and energy usage under various modes of operation. This study compares the performance of the plant in GSP operation with full refrigeration versus partial refrigeration during plant start-up. Comparison is made between plant data and process simulation. The packing hydraulics is studied and compared to vendor hydraulic models. The packing mass transfer efficiency is compared to hydraulic load parameters, separation factor and the stripping factor.

A novel kind of HIGEE-Rotating Zigzag Bed

Wang, G. Q., Xu, Z. C., Yu, Y. L., Ji, J. B.*

Institute of Chemical Engineering and Materials Science, Zhejiang University of Technology, Hangzhou, Zhejiang 310014, Peoples Republic of China.

Abstract

As a new kind of rotating equipment, rotating zigzag bed (RZB) is structurally unique and is of many superior features. RZB is characterized by a rotor coaxially combining a rotational disc with an immobile one. Two series of concentric circular sheets are fixed on the rotational disc and immobile disc respectively, one series as rotational baffles and the other immobile baffles. These baffles are assembled together with each alternate sheets staggered in a vertical direction by about 15mm with respect to adjacent sheets. This arrangement of the baffle sheets provides zigzag flow channels for gas and liquid phase. Compared with conventional rotating packed bed (RPB), RZB could do without liquid distributor and reduce one dynamic-seal, especially easily accomplish intermediate feed in distillation process. RZB also could effectively increase the contact time of gas and liquid phase, thus causing enhancement in mass transfer. In this paper, RZB and its hydrodynamic and mass transfer performance as well as applications were presented.

Keywords

Rotating zigzag bed, rotor, baffle, mass transfer

Background

Up to now, we have witnessed a number of important new developments in chemical industries, which go beyond our traditional chemical engineering and potentially lead to clean, compact, safe, energy-efficient, and environment-benign sustainable processes. These developments were usually termed process intensification, which consists of the development of novel apparatus and techniques that, compared to those commonly used today, are expected to bring dramatic improvements in manufacturing and processing, substantially decreasing equipment-size/production-capacity ratio, energy consumption, or waste production, and ultimately resulting in cheaper, sustainable technologies [1]. Rotating packed bed (RPB), as a result of high gravity (HIGEE) technology, which was invented by New Science Group in Imperial Chemical Industries (London) [2, 3], has developed into one of the most promising branches of process intensification. The core of HIGEE technology is replacing the

* To whom correspondence should be addressed. Tel.: +86 571 88320053. Fax: +86 571 88320053. E-mail: jjb@zjut.edu.cn

gravitational field by a high centrifugal filed (typically 1000g) achieved by rotating a specially shaped rigid bed, usually a disk with an eye in the center. In this way, momentum and heat transfer as well as mass transfer can be tremendously intensified. The rotating-bed equipments, originally dedicated to separation processes such as distillation and absorption, also can be utilized for mass-transfer limited reacting systems. In addition, it potentially can be applied not only to gas-liquid systems, but also to multi-phase systems, e.g. gas-liquid-solid systems.

Fig.1 shows a simplified schematic of a typical rotating packed bed with a vertical axis. The rotor is an annular, cylindrical packed bed housed in a casing and driven by a motor. Liquid is fed through a stationary distributor located at the eye of the rotor as jets or droplets onto the inner periphery of the packing. The liquid attaches onto the packing at the inner periphery and flows over the packing radially outward because of the centrifugal force as thin films, rivulets and droplets, or films covering the pores of the packing depending on the type of packing, the gas/liquid flow rates, and the angular velocity. The liquid leaves from the packing as a shower of droplets, which is collected on the casing wall and outflows at the bottom of the casing. Gas/vapor is introduced into the casing first and then enters into the packing at the outer periphery and is forced to flow radially inward countercurrent to the liquid by pressure driving force. The gas/vapor leaves the packing at the eye of the rotor through the outlet pipe. The mass transfer mainly takes place in the packed rotor. However, perhaps there is a considerable amount of mass transfer taking place between the rotor and the casing [4].

The theoretical basis of HIGEE technology is Sherwood flooding correlation [5], which explains the fluid dynamic performance of packing for distillation and absorption columns. In the Sherwood flooding correlation in form of chart, for a specific system and specific gas/vapor and liquid flowrates, the abscissa, *i.e.* $(u_g^2/g)(a_p/\varepsilon^3)(\rho_g/\rho_L)\mu_L^{0.2}$, is fixed. Based on the assumption that this correlation holds good at high acceleration, we can conclude that if the applied acceleration, g, is increased, the gas velocity, u_g, and the specific area of the packing, a_p, can be increased, thus increasing hydraulic capacity and improving mass transfer performance.

The substitution of centrifugal force for gravitational force in HIGEE could bring many benefits as follows:

- Very high volumetric mass transfer coefficients could be achieved, which would allow the physical size of the equipment to be much smaller and, thus reduce capital and operating costs.
- Higher gas and liquid flow rates could be used, thus the tendency to flood would be reduced.
- Larger driving force of liquid flow due to higher rotational speed could allow to process systems containing non-Newtonian or very viscous Newtonian fluids (such as stripping monomers or solvents from polymers).
- Micromixing, i.e. mixing at molecular scale, could be extremely intensified, which is particularly advantageous to preparation of nanoparticles.

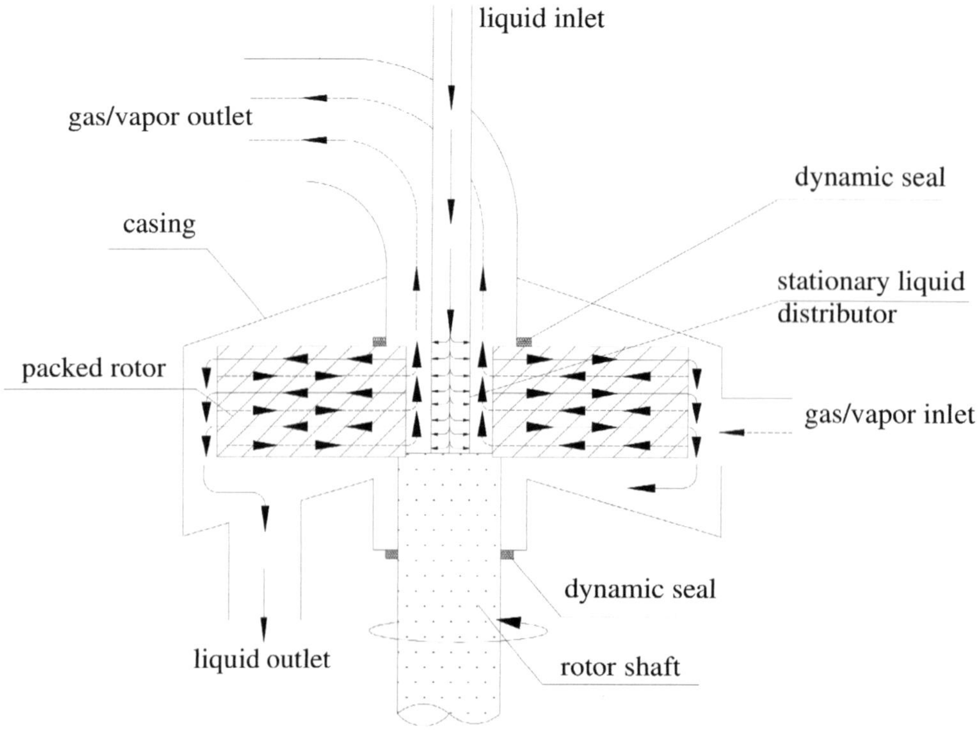

Fig.1 A simplified schematic of typical RPB with a vertical axis.

- The liquid film is thinner and the liquid inventories are smaller, which favors processing precious and poisonous materials.
- The liquid residence time could be dramatically shortened due to smaller size of equipments and higher flow rates, which could avoid the decomposition of heat-sensitive materials in process industries.
- The height and size of equipments could be reduced, which is convenient to installation, troubleshooting and maintenance of facilities.
- The rotor is self-cleaning and does not subject to plugging, which benefits to treatments of fouling and solid-containing systems.
- The HIGEE unit will be unaffected by moderate disturbance in its orientation, which relaxes a major limitation on the use of conventional columns in barge or ship-mounted units where only very small deviations from the vertical are acceptable.

In spite of these advantages, RPB has been little used commercially, mostly in separation operations, probably because of inadequate understanding of the contact mechanism of two phases in RPB and practitioners' preference for conventional packed beds without moving parts. In fact, RPB could be applied to all mass-transfer-limited processes such as gas-liquid system with high reaction rates.

The main difference between all kinds of high gravity rotating beds is structures of the rotor. Now there mainly exist five types of rotors in open literature, *i.e.* packing bed, disk plate [6], helical [7], multi-staged spraying [8] and split packing[9,10]. However, the RPB still is the commonly used compared to the other ones. The zigzag rotating bed (RZB) was invented and patented by Ji *et al*[11, 12]. The aim of this paper is to introduce RZB and its performance and applications.

Introduction

RZB is a new kind of HIGEE characterized by a rotor coupling a rotational disc with an immobile one. One series of concentric circular sheets as rotational baffles, the upper part of which is perforated, is fixed on the lower (rotational) disc driven by a motor through a vertical shaft. The other series of concentric circular sheets as immobile baffles is mounted on the upper (immobile) disc. The two groups of baffles are arranged oppositely and alternately, as shown in Fig.2. The structure of the rotational and immobile baffles was presented in Fig.3. The clearance between the rotational baffle and the upper disc as well as between immobile baffle and the lower disc offers flow channels for the gas and liquid phase. As conventional RPB, in the RZB the gas is introduced into the casing and flows inward radially driven by pressure difference. The liquid is fed at the center of upper disc flowing outward countercurrent to gas due to centrifugal force. During contacting process the gas phase is continuous and liquid phase is dispersive.

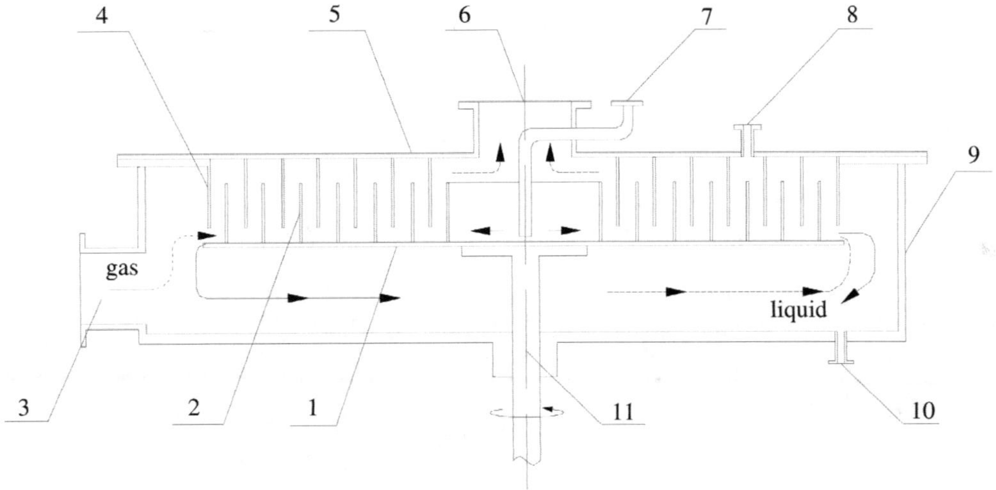

1-rotational disc 2-rotational baffle 3-gas inlet 4-immobile baffle 5-immobile disc 6-gas outlet 7-liquid inlet 8-Intermediate feed 9-rotor casing 10-liquid outlet 11-rotating shaft

Fig.2 Simplified sketch of the rotor of RZB

Compared with conventional RPB, RZB is featured by following advantages:

- The upper disc of the rotor is immobile and the lower disc is rotational, therefore the upper dynamic-seal used in RPB (as shown in Fig. 1) could be replaced by a static-seal. This way facilitates the manufacture, installation and maintenance of the equipment. The safety and reliability was improved.

- Intermediate feed for distillation column could easily be accomplished by installing an inlet at the upper disc because the upper disc is immobile. Thus it is possible for only one unit of RZB to perform a continuous distillation duty. Contrastively, two units of RPB are necessary if applied to distillation services due to incapability of feeding the unit at some point equivalent to a mid-plate position because the whole rotor is rotating. The arrangements of RPB and RZB for distillation process are shown in Fig.4.
- The contacting elements of gas and liquid phase in RZB are baffles, through which a good liquid distribution could be obtained because the liquid is thrown by centrifugal force from the rotational baffles and impinges on the immobile baffles resulting in very fine droplets. Therefore RZB does not require initial liquid distributor.
- In RZB, the gas/liquid phase can be considered as continuous/dispersive zigzag flow through the rotational and immobile baffles, so the total length of flow channel and the residence time of gas and liquid phase are extended compared with RPB, which results in an increase of mass transfer capacity.
- For RZB, it is easy to install multiple rotors in one casing, thus a fold increase of theoretical plate number in a single unit could be obtained. Fig.5 shows a schematic diagram of multi-rotors of RZB with a liquid guide pipe in the eye of each immobile disc of rotor. However, as far as RPB is concerned, one casing only can hold one rotor.

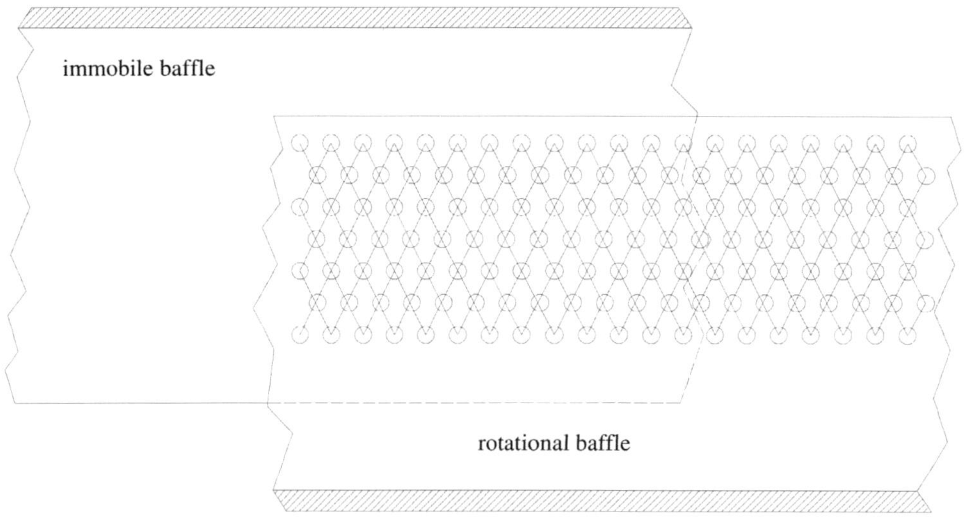

Fig.3 Structure of rotational and immobile baffles

Mass transfer mechanism

In the interior of rotor, the gas flows along a zigzag path and the liquid experiences multiple dispersion- agglomeration recycles. The gas fed by the gas inlet in the casing enters into the rotor at the periphery of the rotor, and then flows radially inward in a staggered way through the space between the rotational and immobile baffles and finally discharges through the gas outlet pipe located at the eye of immobile disc. The perforated rotational baffles allow gas to pass and prevent liquid from back-mixing. The liquid introduced by inlet at the center of the immobile disc is first dispersed in the form of fine droplets by rotational baffles with high speed. The droplets are intercepted by immobile baffles, then agglomerate to form liquid again falling on the rotational disc. The liquid flows radially outward to the next rotational baffle and is dispersed again. During this process, impingement and spray of the liquid in the rotor results in fast-renewed gas-liquid interface with very large specific surface area available for mass transfer.

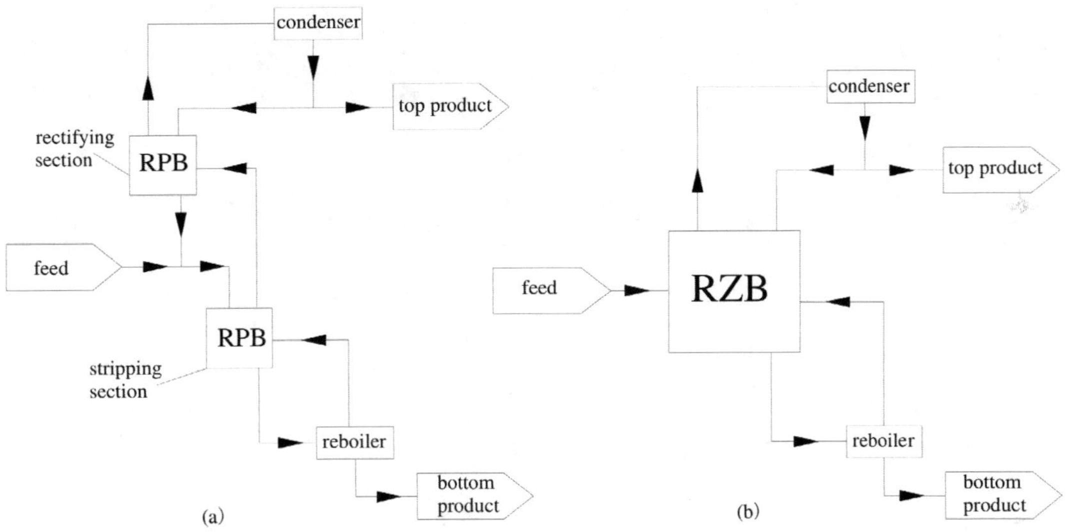

Fig.4 HIGEE for distillation setup (a) RPB (b) RZB

In the rotor, the gas and liquid phase contacts countercurrently on the whole. However, two-phase contact process in the annular space between the rotational and immobile baffles can be further divided into two steps. The first step is crosscurrent contact of two phases when the liquid is thrown by the rotational baffles. The second step is countercurrent contact of two phases when the liquid falls down along the immobile baffles. The contact process of gas and liquid in RZB was shown in Fig.6. In the second step, the surfaces of falling liquid renew rapidly due to rotary motion and impingement by liquid droplets, and the interfacial mass transfer rate is very high. The contact of gas and liquid phase in one annular space of the rotor is equivalent to a small wetted-wall column and RZB can be considered as wetted-wall cluster in centrifugal field. Because centrifugal force is much greater than gravity, the turbulence of gas-liquid phase in wetted-wall cluster in centrifugal field is much greater than that of wetted-wall column in gravitational field. Therefore the inter-phase transport processes are extremely intensified, and the interfacial mass transfer rate is greatly enhanced.

As mentioned above, it is impossible for RPB to coaxially install multiple rotors in one casing, because the liquid from the periphery of the upper rotor cannot return to the eye of the lower rotor due to rotation of the lower rotor. But for RZB it is not the case. The liquid leaving from the periphery of upper rotor can be easily collected and directed by the guide pipes into the eye of the lower rotor because the upper disc of the lower rotor is immobile. The gas emerging from the eye of the lower rotor enters into the upper rotor from its outside edge, and flows radially inward until the eye of the upper rotor. In this way multi-rotor in one casing is easily accomplished, which makes the equipment more compact. The flow paths of gas and liquid in multi-rotors RZB are illustrated in Fig.2 and 5.

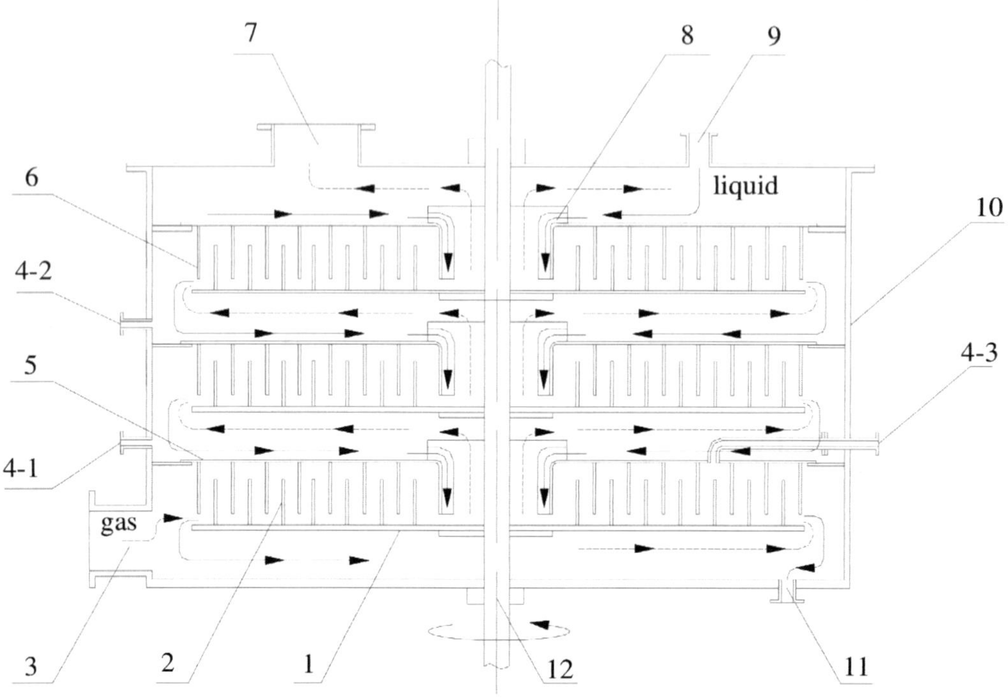

1-rotational disc 2-rotational baffle 3-gas inlet 4-1,2,3-liquid inlet 5-immobile disc 6-immobile baffle 7-gas outlet 8-guide pipe 9-reflux pipe 10-rotor casing 11-liquid outlet 12-rotating shaft

Fig.5 Simplified sketch of RZB with three-rotors in a casing

Hydrodynamic and mass transfer performance

The hydrodynamic experiments were carried out in a RZB using air-water system. The rotor is 600/150 mm in outer-/inner-diameter and 80 mm in height. The casing is 800mm in diameter with an aspect ratio of unit. The gas pressure drop dependence on gas and liquid flowrate as well as the rotational speed was given in Fig.7-9. From these figures it can be seen that the gas pressure drop obviously increases with gas flowrate at a constant rotational speed and linearly increases with rotational speed for different gas flowrate. The dependence of gas pressure drop on liquid flowrate is a little complicated at a constant gas flowrate. At a lower liquid flowrate, the pressure drop decreases with liquid flowrate, but at a higher liquid flowrate the pressure drop almost keeps a constant. A possible reason is that at a lower liquid flowrate, the actual gas flowrate is almost constant because of small flow area of liquid. The vacuum formed by liquid droplets that were thrown by centrifugal force facilitates the gas flow inward and results in reduction of flow resistance. However, at a higher liquid flowrate, reduction of flow resistance is offset by increase of gas pressure loss, which results from the increase of gas actual flowrate due to the decrease of flow areas.

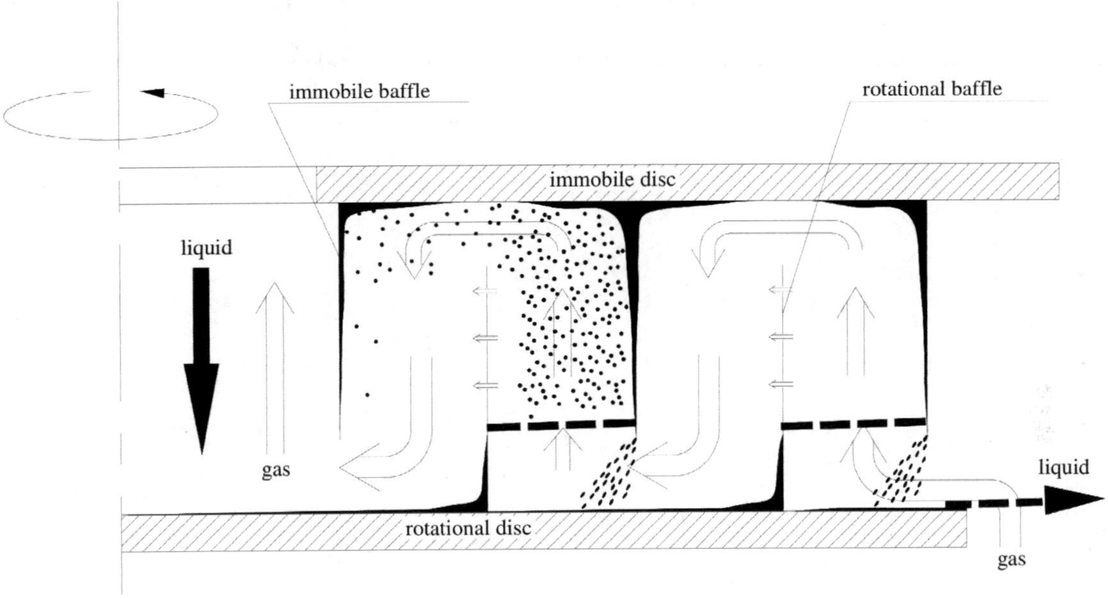

Fig.6 The contact process of gas and liquid in RZB

The mass transfer experiments were conducted in a RZB with two rotors in one casing using ethanol-water system. The rotor is 630mm in outer diameter, 80mm in height and the casing is 800mm in diameter, 550mm in height. There are nine rotational and ten immobile baffles for the upper rotor, and seven and eight for the lower. The investigation is focused on the effects of reflux and rotational speed on the theoretical plate number, which are given in Fig.10. It is indicated that with the reflux increase, the mass transfer performance first improves and then levels off. The rotational speed of rotors has an obvious effect on the mass transfer efficiency. For revolution lower than 1000rpm, the theoretical plate number increases with rotational speed and then keeps almost constant, which shows the perfect contact of gas and liquid phase at this time. On the basis of effective radial flow length, the separation ability of RZB is 20 theoretical plates per meter with 1000rpm and 600l/h reflux.

Industrial application

RZB have been successfully applied in many separation services, especially continuous distillation, e. g. methanol and ethanol recovery.

As for methanol-water system, three rotors with outer diameter 750mm are installed in one casing that is 830 mm in diameter and 800 mm in height. With rotating speed of 1000 rpm and reflux ratio of 1.5 and methanol mass concentration in feed of 70%, the concentration of the top product is larger than 99.7% and the concentration of the bottom residual liquid is less than 0.5%, with product output: 500 kg/h. The total theoretical plates are 16 with rectifying section 11.6.

For ethanol-water system, two rotors with outer diameter 630mm are installed in one casing that is 700 mm in diameter and 600 mm in height. With rotating speed of 1000 rpm and reflux ratio of 2.5, if the mass concentration and flowrate of feed is 45% (ethanol) and 470 kg/h respectively, the concentration of the top product is larger than 95% and the concentration of the bottom residual liquid is less than 0.5%.

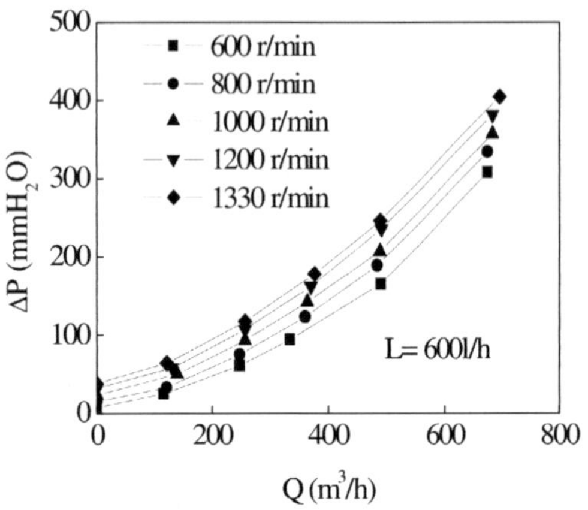

Fig.7 Variation of gas pressure drop with gas flowrate at different rotational speeds

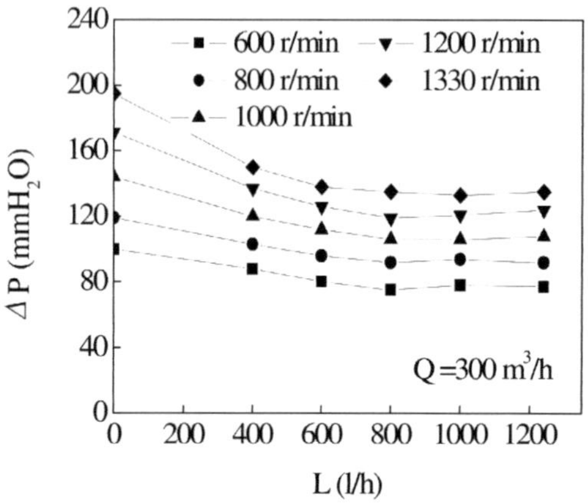

Fig.8 Variation of gas pressure drop with liquid flowrate at different rotational speeds

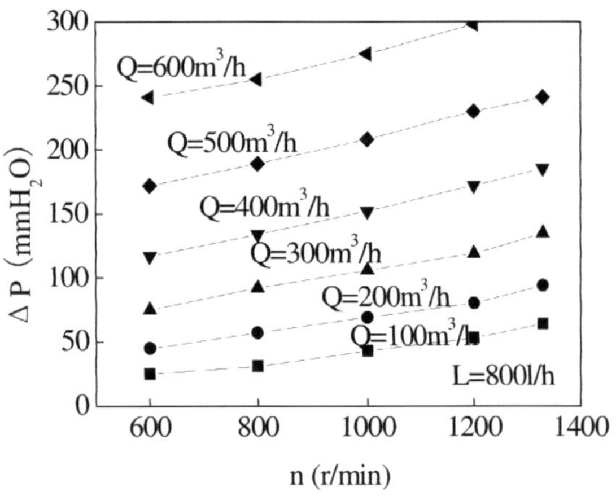

Fig.9 Variation of gas pressure drop with rotational speed at different liquid flowrates

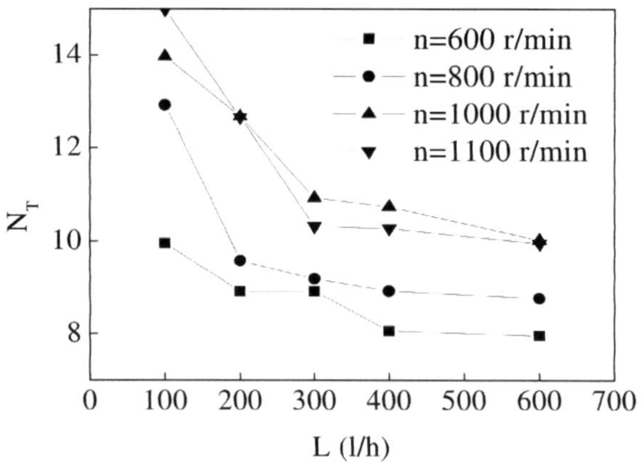

Fig.10 Variation of theoretical plate number with reflux at different rotational speeds

The sizes of RZB and the conventional packed column (PB) are compared for a given separation task, as shown in Table 1. It can be seen that an 1 to 2 orders of reduction in height and volume of equipments can be achieved in RZB and thus it is a kind of gas-liquid contactor for process intensification. RZB is a new kind of HIGEE equipment that could be industrially applied in continuous distillation processes.

Table1 comparison of the size of RZB and PB for a given separation task

	Ethanol-water		Methanol-water	
	RZB	PB	RZB	PB
diameter/ m	0.80	0.4	0.83	0.6
total height/ m	0.55	9.0	0.80	11.0
volume/ m^3	0.276	1.13	0.433	3.11
height ratio/ %	6.1		7.3	
volume ratio/ %	24.4		13.9	

Further work

As a novel kind of equipment, although RZB have been successfully utilized in chemical industries, further experimental as well as theoretical investigations are much wanting. More work needs to be done to improve the predictions and understand the fluid flow and mass transfer processes on a rigorous, fundamental basis. Another piece of work is establishment of a reliable procedure which is premise of design and optimization of its physical dimensions.

Now RZB's applications are mainly focused on continuous distillation, but it is general equipment applicable to all mass transfer or mass-transfer controlled processes. So to speed its industrialization and extend its application area, such as absorption, stripping, gas-liquid reaction system, is of practical importance and should be explored.

Acknowledgements

This Project was supported by Zhejiang Provincial Natural Science Foundation of China through grant Y406239, which is gratefully acknowledged.

Reference

1. Stankiewicz, A.; Moulijn, J. A. Process intensification: transforming chemical engineering. *Chem. Eng. Prog*. **2000**, 96, 22.

2. Ramshaw, C.; Mallinson, R. H. Mass transfer apparatus and its use. European patent. EP0002568,

1979.

3. Ramshaw, C.; Mallinson, R. H. Mass Transfer Process. United States patent. US4383255, **1981**.

4. Chen, Y. S., Lin, C. C., Liu, H. S. Mass Transfer in a Rotating Packed Bed with Various Radii of the Bed. *Ind. Eng. Chem. Res.*, 2005, 44, 7868.

5. Sherwood, T. K.; Shipley, G. H.; Holloway, F. A. L. Flooding Velocities in Packed Columns. *Ind. Eng. Chem.* **1938**, 30, 765.

6. Jian, Q. F.; Deng, X. H.; Deng, S. J. Experimental investigation of gas flow resistance and gas-liquid mass-transfer performance in super-gravity force rotating packed bed with wave form disk packing. *Chem. React. Eng. Technol. (China)* **1998**, 14, 42.

7. Chen, Z. Q.; Xiong, S. X.; Wu, J. G. Helical rotating absorber. *J. Chem. Ind. Eng. (China)* **1995**, 46, 388.

8. Pan, Z. Q.; Zhang, Y. J.; Deng, X. H., *et al.* Experimental investigation into mass transfer between liquid and gas in multi-staged spraying rotating packed bed. *J. South China Univ. Technol. (China)* **2006**, 34, 67.

9. Chandra, A.; Goswami, P. S.; Rao, D. P. Characteristics of Flow in a Rotating Packed Bed (HIGEE) with Split Packing. *Ind. Eng. Chem. Res.*, **2005**, 44, 4051.

10. Reddy, K. J.; Gupta, A.; Rao, D. P.; Rama, O. P. Process intensification in a HIGEE with split packing. *Ind. Eng. Chem. Res.*, **2006**, 45, 4270.

11. Ji, J. B.; Wang, L. H.; Xu, Z. C. *et al.* Equipment of zigzag high-gravity rotating beds. China patent. CN01134321.4, **2004**.

12. Ji, J. B.; Wang, L. H.; Xu, Z.C. *et al.* Equipment of multi-rotors zigzag high-gravity rotating beds. China patent. CN200520100685.3, **2006**.

A New Approach to Intensify the Structured Packing

Weiyang Fei, Suojuan Luo, and Xinyue Song. Dept. of Chem. Eng, Tsinghua University, Beijing, 100084, China

The structure packing has been widely applied to distillation and absorption for chemical based industrial for decades. However, its capacity and efficiency haven't been improved too much while the demands on high efficient column internals are increasing in the last 20 years. Therefore, the study on new approaches to intensify the packing becomes an active research topic which can bring remarkable economical benefits.

Gas and liquid flow behavior in corrugated structure packing was studied by CFD simulation. It is shown from the results that the collision of cross gas streams causes main part of packing bed pressure drop which makes no contribution on gas-liquid mass-transfer. Furthermore, the liquid deflective flow on traditional corrugation packing surface was notable. Therefore, effective mass-transfer area was reduced.

A new type corrugated packing has been developed based on the CFD simulation and theoretical analysis. It is clear from our studies that reducing the Channel opening angle (Crimp angle) at same specific surface area can reduce the gas pressure drop and control the liquid deflective flow. As a result, the effective mass-transfer area and the inter-phase mass transfer efficiency increased.

The characteristics of the new packing have been studied in a Φ50mm distillation column. It was shown from the results that the pressure drop of the new packing reduced about 30% and mass-transfer efficiency increased about 15% compared with the Mellapak with the same special area.

Thermodynamics Insights: Solving Distillation Mysteries Using a Simulation Lens

Kenneth R. Cox, Rice University, 6100 Main Street, Houston, TX 77005-1892

All popular process simulation programs contain thermodynamic packages that can contribute to design and troubleshooting of separation systems involving distillation. Still, there is nothing completely automatic about selecting an appropriate physical property package containing the best parameters for the job at hand. If we choose poorly, the method may not yield the accuracy we require for design and troubleshooting challenging systems. Are there heuristic approaches and simple physical concepts that can aid us in this important selection process?

Properly-selected physical property packages enable the effective use of modern simulation tools. Once enabled, the engineer can use these tools to for troubleshooting columns, selecting entrainers, predicting azeotropes and scale formation, and many other design and operations decisions. Accurate property prediction is essential to the success of all of these tasks.

In this talk we will present several approaches a practicing engineer can use better leverage the tools available in process simulation packages. This will include heuristics and shortcut methods that will complement the sophisticated tools available in these packages.

The Final Step to Success – Tower Internals Inspection Part 1

Ed Grave and Paul Tanaka. ExxonMobil Upstream Research Company, P.O. Box 2189, URC-URC-S354, Houston, TX 77252-2189

An effective inspection program must be implemented early in the Engineering/Procurement/Construction phase and defined during Front End Engineering Design. Suppliers, contractors, and subcontractors must be part of the overall program, but most importantly, owners and operators must take a more proactive approach to field inspections. The most undesirable design mistakes are those that have not been identified during the drawing review stage. Subcontractors lack expertise in fractionation and separation to catch these design errors in the field.

Successful startup and operation of fractionation towers rely on quality design, fabrication, and installation. As a project proceeds from design towards installation, the cost, complexity, and time to implement any required corrective action substantially increases while the number of practical corrective options decreases. Although identifying and correcting issues is best done during the design phase, not all issues will be discovered and rectified prior to fabrication and installation. Shop and field inspections provide the last best hope of identifying and correcting any issues to ensure successful startup and operation of fractionation towers.

Shop inspections are paramount to discovering and correcting major design and fabrication issues that will affect internals installation and tower operation. Major issues cannot be resolved in the field due to the lack of adequately-equipped facilities. Field inspections are vital to ensure that the assembled tower will function according to the intended design. However, short project schedules, lack of resources, and inexperienced personnel are a recipe for poor installation and assembly of internals.

The purpose of this presentation is to describe key aspects of an effective and practical inspection program and recent examples of the types of issues that can be encountered.

Debottlenecking of Hexane Distillation

Gerhard Ruffert, Dr. Bausa, and Di Halleneberger. Bayer Technology Services GmbH, Leverkusen, Germany

Hexane distillations seam to be a very common task in chemical and petrochemical industries. Nevertheless, the special behaviour of mixtures to be separated may lead to surprising effects inside the column. The presentation represents a case story of a hexane distillation with a 3 step debottlenecking development. By means of design numbers and Gamma-Scan diagrams the performance of the column is analysed and discussed during the 3 troubleshooting and revamp stages.

The Art of Using Simulator Testing and Data for Design of Pressured Industrial Applications

Mark W. Pilling, Sulzer Chemtech USA, Inc., 4019 S. Jackson Ave., Tulsa, OK 74107, Ansor Gaebler, Sulzer Chemtech Switzerland, Sulzer-Allee 48, P.O. Box 65, Winterthur, 8404, Switzerland, and Marc B. Wehrli, Sulzer Chemtech AG, Sulzer-Allee 48, Winterthur, CH-8404, Switzerland.

Ambient simulation columns are the most common type of test facilities used for the evaluation of distillation internals. This is due to their economical installation and operating costs. The fundamental problem with simulators is that the system properties of the test fluids do not directly match the properties of the actual industrial fluids. Sometimes, the differences are rather small, as with atmospheric columns. Other times, the differences are quite significant, as with high pressure applications. This paper will evaluate how and when ambient simulation test data can be used effectively for evaluation of internals. It will also discuss the pitfalls of relying on such data for the design of internals for poorly suited industrial applications.

Effects of a Trapped Trace Component on Operation of a Wastewater Stripper/Organic Recovery Tower

Bruce S. Holden and Patrick H. Au-Yeung. Core R&D/Process Separations, The Dow Chemical Company, 1319 Building, Midland, MI 48667

Promulgation of Maximum Achievable Control Technology (MACT) regulations for cellulosics manufacturing by the U.S. EPA required the installation of a wastewater stripper in a production unit for cellulose ethers in 2005 to remove low levels of methanol and other light compounds classified as hazardous air pollutants (HAPs) by the regulation. To purify the stripped methanol for potential use in other processes or for outside sale, a rectification section was added to the stripper to hold down water and other heavy impurities. Shortly after startup, an inability to hold water down out of the distillate and a series of significant tower upsets were eventually linked to the presence of a previously unidentified low-concentration intermediate component that became trapped in the tower. Changes to the tower allowed stable tower operation and reduced the amount of water carryover. This paper describes the observed effects of a trapped component on tower operation including damage to tower internals, the vapor-liquid equilibria of the multicomponent system resulting in the trapped component, the control strategy changes that resolved the problem, and the circumstances and pitfalls that caused the problem to remain unidentified prior to startup.

Guidelines for Ethylene Quench Towers

Tim Zygula[1], **Karl Kolmetz**[2], Chee Mum Tham[2], Wai Kiong Ng[3], and Jun H. Chan[2]. (1) Huntsman, 2504 South Grandview, Odessa, TX 79766, (2) GTC Technology LP, Singapore, Singapore, (3) Crystallisation and Particle Science, Institute of Chemical and Engineering Sciences, 1 Pesek Road, Jurong Island, 627833, Singapore

Ethylene quench oil and quench water towers may have the highest failure rates in fractional distillation due to the rigorous service that these tower perform. Almost all of the challenge areas of distillation are concentrated into one column system; high temperature, solids, fouling potential, oxygenates, polymerization potential, heat removal by pump arounds, and a mixture of Hydrogen, Steam, and C1 to C20s. Given this combination, the guidelines for designing this column have to be flexible yet sturdy to handle each of the challenges.

One of the most reliable designs has been the one developed by Timothy M. Zygula and Karl Kolmetz in 1998. These designs have run continually for eight years, with previous designs in these revamped columns having less than one year run length. The authors will examine the history of quench oil and quench water tower's successful and not successful case studies and lessons than can be learned from each of the cases.

Methods for Obtaining Detailed Process Knowledge about Distillation Operations

Lowell Pless, Tracerco, 11005 W. Fairmont Pkwy, La Porte, TX 77571

For distillation and separation processes Chemical Engineers have historically relied upon plant process measurements such as flow rates, temperatures, and pressures and model or simulation results for data with which to solve troubleshooting or re-design problems. An array of on-line diagnostic services are available that can provide additional data offering real-time information on how pieces of process equipment are actually operating.

The most common application is gamma scanning. This test is primarily applied, but not necessarily limited to, distillation or separation columns. Gamma scans provide a density profile of the internal process of operating distillation columns and other process vessels. The density profile can be used to diagnose the hydraulic operating conditions of mass transfer devices such as damage to internals, flooding, degree of entrainment or weeping, liquid levels on trays and distributors, liquid distribution through packed beds, etc. The presenter will show case studies where scanning revealed vital process information that helped solve an operating problem, or helped make a revamp/re-design successful.

Additional derivatives from gamma scanning, CAT-scans and RapidScan, will be demonstrated. The CAT-Scan technique is a specialized horizontal gamma scan used to generate a topographic profile of the internal cross-sectional density of process equipment. This profile is useful for the detailed study of liquid flow distribution through packed columns as well as having applications beyond distillation columns. RapidScan is a new hardware system that allows for extremely accurate and repeatable gamma scan data acquisition in cases where comparison of historical data is important.

Radioactive tracers are used on operating processes to acquire real-world measurements of process variables that otherwise could only be calculated or estimated. Process variables that can be directly measured with radioactive tracers include flow rates, flow distribution, residence times, mixing characteristics, etc. Case studies of tracer applications applied specifically to distillation operations such as entrainment studies and leak tests will be discussed.

The Final Step to Success – Tower Internals Inspection Part 2

Ed Grave and Paul Tanaka. ExxonMobil Upstream Research Company, P.O. Box 2189, URC-URC-S354, Houston, TX 77252-2189

An effective inspection program must be implemented early in the Engineering/Procurement/Construction phase and defined during Front End Engineering Design. Suppliers, contractors, and subcontractors must be part of the overall program, but most importantly, owners and operators must take a more proactive approach to field inspections. The most undesirable design mistakes are those that have not been identified during the drawing review stage. Subcontractors lack expertise in fractionation and separation to catch these design errors in the field.

Successful startup and operation of fractionation towers rely on quality design, fabrication, and installation. As a project proceeds from design towards installation, the cost, complexity, and time to implement any required corrective action substantially increases while the number of practical corrective options decreases. Although identifying and correcting issues is best done during the design phase, not all issues will be discovered and rectified prior to fabrication and installation. Shop and field inspections provide the last best hope of identifying and correcting any issues to ensure successful startup and operation of fractionation towers.

Shop inspections are paramount to discovering and correcting major design and fabrication issues that will affect internals installation and tower operation. Major issues cannot be resolved in the field due to the lack of adequately-equipped facilities. Field inspections are vital to ensure that the assembled tower will function according to the intended design. However, short project schedules, lack of resources, and inexperienced personnel are a recipe for poor installation and assembly of internals.

The purpose of this presentation is to describe key aspects of an effective and practical inspection program and recent examples of the types of issues that can be encountered.

A Design Review of Steam Stripping Columns for Wastewater Service

Tim Zygula, Huntsman, 2504 South Grandview, Odessa, TX 79766

In recent years the role of steam strippers in industry has become an integral part of plant operations. The goal when designing these types of columns is to have reliable operation for as long as possible. In order to achieve this goal the designs of these columns have to be carefully considered. The authors will detail the methodology used when designing steam strippers for wastewater service. Some of the topics that will be covered by the authors are:

1. Steam Stripper Design Fundamentals.

2. The importance Process Chemistry has on the column design.

3. Operation of a steam stripper.

4. How to select internals for a steam stripper column.

The authors will present an example of a steam stripper design.

Ethyl Lactate Production Using Semicontinuous Distillation with Reaction in an Auxiliary Vessel and Pervaporation

Thomas A. Adams II and Warren D Seider. Chemical and Biomolecular Engineering, University of Pennsylvania, 352 Towne Building, Philadelphia, PA 19104

At the 2005 Spring AIChE Distillation Symposium, we introduced a novel concept that combined semicontinuous distillation with reaction in a middle vessel (SDRMV) to produce 2,4-dimethyl-1,3-dioxolane in a forced cyclic process [1]. At the Cincinnati Meeting of AIChE (November 2005), we showed that this system was also economically preferred to traditional batch and continuous alternatives for a wide range of production rates [2]. Now, we present a new application of integrated semicontinuous distillation, reaction, and pervaporation for the production of ethyl lactate from ethanol and lactic acid.

Semicontinuous distillation uses one or more middle vessels integrated with a distillation column in a forced cycle consisting of a sequence of operating modes. With this method, three or more chemical species can be separated with just one column at purities higher than a sidedraw would permit. Unlike divided wall columns, only a single shell is needed, and so the column is simple and control is easier. Reaction can take place in the middle vessel or in an auxiliary vessel, which when integrated with the distillation column, allows for either the simultaneous removal of products or the recycling of unreacted reagents to help push the reaction forward. This method is sometimes preferable to reactive distillation (reaction on the trays of the column) for exothermic processes or where the bubble-point temperatures inside the column do not favor fast reactions or high yields.

In this work, equimolar amounts of ethanol and lactic acid are reacted to chemical equilibrium in a CSTR to form equimolar amounts of water and ethyl lactate, which is the desired product. The reaction is exothermic and reversible with a well-known azeotrope existing between ethanol and water. Once the reaction approaches chemical equilibrium, the contents of the reactor (containing all four species) are moved to a middle vessel that is integrated with a distillation column, and the CSTR is recharged. The middle vessel feeds to and receives a sidedraw from the column simultaneously.

Throughout the cycle, ethanol and water are collected in the distillate and are separated with a pervaporation unit utilizing a hydrophilic membrane. This permits a low-energy separation without complications due to the azeotrope. The water-rich permeate is collected and removed from the system, while the ethanol-rich retentate is recycled to the CSTR, pushing the reaction forward. Lactic acid is collected in the bottoms product of the column and recycled to the CSTR at decreasing flow rates, eventually approaching zero flow. At this point, the process resembles a fed-batch distillation with a sidedraw. Any ethyl lactate product collected in either stream is kept within the process since the streams are recycled to the CSTR, thus achieving zero product loss. Since water, ethanol, and lactic acid are removed from the middle vessel during this process, the middle vessel becomes highly concentrated with ethyl lactate approaching the end of the cycle, when the contents are collected as product.

The flow rates and compositions of the feed, distillate, and bottoms vary throughout the process and require a special model-based control system to maintain purity constraints and prevent the column from failing due to flooding or weeping. A sequence of operating modes are needed to ensure seamless operation from mode-to-mode within the cycles, so that unlike batch distillation, there are no startup or cooldown phases. Thus, the process is semicontinuous in nature.

The process is simulated rigorously through integration of the dynamic MESH equations for the column, mass and energy balance equations for the tanks, and empirical reaction equations for the CSTR. The pervaporation model is derived from a mass-transfer-based bulk flow model for continuous systems and is adapted for use in a dynamic environment. Physical properties are modeled rigorously through calls to Aspen Properties 2004.1.

The results demonstrate the feasibility of using semicontinuous distillation to separate a four species mixture into three mixtures of controllable purity using a single distillation column. This requires less capital expenses than using alternative batch or continuous processes. Additionally, when integrated with a reaction unit and pervaporation unit, the system can help increase reaction yield and circumvent simple distillation boundaries associated with the azeotropes. Furthermore, the operation of the column without startup or cooldown phases between the batches permits a higher rate of production to be achieved for processes of comparable size. Thus, the process is a good candidate for use at intermediate production rates between those typically associated with batch and continuous processes.

1. Adams TA, Seider WD. A novel concept: Semicontinuous reactive distillation. 2005 AIChE Spring Nat'l. Meet., 2005. Atlanta, GA, United States: American Institute of Chemical Engineers, New York, NY 10016-5991, United States.

2. Adams TA, Seider WD. Semicontinuous Reactive Distillation for Specialty Chemicals Production: Economic Comparison with Batch and Continuous Processing. 2005 AIChE Annual Meet., 2005. Cincinnati, OH, United States: American Institute of Chemical Engineers, New York, NY 10016-5991, United States.

A Fresh Look at Underwood's Method

Angelo Lucia, University of Rhode Island, Dept of Chem. Engineering, Kingston, RI 02881 and **Amit Amale**, Chemical Engineering, University of Rhode Island, Dept of Chem. Engineering, Kingston, RI 02881.

Energy consumption in the chemical and petrochemical industries is dominated by distillation. However, distillation is still the most versatile means of separation and will continue to be used in some capacity to address a wide variety of separation needs. With the rising cost of energy, there is renewed interest in the design and/or retrofit of energy efficient processes so that the end products are more cost effective.

Despite its limitations, Underwood's method remains a widely used tool for determining minimum energy needs in distillation – largely because it is straightforward and gives good results in many situations. Recently our group has introduced a new methodology – the concept of the shortest stripping line – to determine minimum energy requirements for multicomponent separations involving ideal, nonideal, azeotropic and reactive mixtures. This new methodology can also determine minimum energy consumption for multi-unit processes like hybrid separations (e.g., distillation plus extraction) or reaction/separation/recycle.

This presentation centers on Underwood's method and its relationship to the concept of shortest stripping lines. In particular, the shortest stripping line approach is further extended for analyzing the sensitivities of column composition profiles, and in turn energy requirements, to distillation product compositions. Since Underwood's method is a group method and does not involve rigorous mass balance calculations, it cannot be used for this type of analysis. It is also well known that methods like the boundary value methods of Doherty and co-workers can show strong sensitivy to product compositions – especially with regard to components present in trace amounts. As a result, feasibility difficulties can arise and marked differences in calculated energy use can result in boundary value methods. It is thereofre desirable to develop a method that finds the best product compositions so that the desired separation is met and minimum energy is used.

In this work, a methodology that combines a robust optimization technique with the shortest stripping line approach is proposed to address the problem of sensitivity of column composition profiles and energy requirements to specified product compositions. The overall strategy, which can be applied to mixtures with any number of components, involves two levels of optimization. In the inner loop, product compositions consistent with the desired separation are used as optimization variables to ensure feasibility while the stripping or reboil ratio is used as the optimization variable in the outer loop to minimize the associated energy requirements. A terrain-following method, which has also been developed in our group, is used for solving the resulting nonlinear programming problems. This two-level optimization methodology is applied to a variety of multicomponent distillations and detailed numerical results are presented to illustrate the robustness of our approach. These numerical results clearly show that our two-level optimization approach can be used to reliably find the best combination of components in the products so that the desired separation requires minimum energy and to quantify the relative sensitivities of column profiles to product compositions. Our numerical results are also compared and contrasted to results from Underwood's method.

Mass Transfer - Thinking beyond Theoretical Stages

Kenneth Graf, Koch-Glitsch, 4111 E 37th Street N, Wichita, KS 67220 and Jeff DeGarmo, R&D, Koch-Glitsch, 4111 E 37th Street N, Wichita, KS 67220.

Equilibrium stage calculations have become nearly universal calculation tools for mass transfer columns contacting vapor and liquid in countercurrent flow. Whether the service is absorption, stripping, distillation, fractionation, quench, or evaporation, equilibrium stage models make mass and heat balances easy and complete.

In some columns, equilibrium or theoretical stages are a good model of what the vapor/liquid contacting devices provide. But theoretical stages are often a poor model of what trays and packing provide. In such cases, equilibrium stage calculations are still highly useful for establishing heat and mass balances, for finding stream conditions, and for estimating stream properties. But then the engineer must realize that correlating tray or packing performance to theoretical stages is deceptive, and he or she should not force-fit a design tray efficiency or packing HETP to a column design. Many times, a force-fit tray efficiency of 30% or a packing HETP three times the standard HETP is interpreted as equipment that doesn't work very well. In reality, it is the reliance on the theoretical stage model that doesn't work very well.

This paper discusses columns in which theoretical stages are a poor representation of tray or packing performance, and it provides guidelines to help an engineer find a more consistent and logical model.

Using the Analysis of the Asymptotic Behaviour of Column Profiles to Define the Overall Distillation Behaviour in Transformed Residue Curves and Column Profiles

Brendon Hausberger[1], Diane Hildebrandt[2], and David Glasser[2]. (1) Centre of Material and Process Synthesis, School of Chemical and Metallurgical Engineering, University of the Witwatersrand, Johannesburg, 1 Jan Smut Avenue, Johannesburg, South Africa, (2) COMPS, School of Chemical and Metallurgical Engineering, University of the Witwatersrand, Johannesburg, 1 Jan Smut Avenue, Johannesburg, South Africa

Previous work by Hausberger et al. demonstrated the asymptotic behaviour of column profiles as they move towards their respective pinch points. More recent work by Tapp et al. 2004, has demonstrated that column profiles actually follow the trajectory defined by a simple transform of the original residue curve of the system under examination for their entire length. In this paper we examine the linkage between these areas of research in that the eigenvectors used to define the asymptotic behaviour are also shown to define the directions of the boundaries of the transformed triangles. This is a useful result in that the previous work for transform triangle focused on constant volatility systems which have straight line boundaries as for more complex systems, the behaviour become complex. The asymptotic analysis permits the simple determination of the directions of the apexes of the transformed "triangles". As this information is generic to all transformed triangles terminating on a given point, this provides a powerful insight in to the behaviour that could be expected of a column operating in that transformed section. The paper also provide some practical applications of this tool.

References:

Tapp, M., Holland, S.T., Hildebrandt, D. and Glasser, D. (2004) Column Profile Maps Part A: Derivation and Interpretation; Ind. Eng. Chem. Res., Vol 43, 2, 364-374.

Hausberger, B., Hildebrandt, D., Glasser, D. Asymptotic Behaviour of Distillation Equations. Presented at Chemeca '99, Newcastle, Australia, 1999

Web Page: www.comps.wits.ac.za

Un-Predictables in Distillation Towers - Fouling and Foaming

Simon Xu, Process Technology, Shaw Stone & Webster Inc, 1430 Enclave Pkwy, Houston, TX 77077

Fouling and foaming are frequently at the top of a list of "well known but little understood" problems in distillation towers. The difficulty is its "unpredictable" characteristics which are hard to repeat or correlate in lab experiments or industrial operations. Chemical reactions and surface phenomena in fouling and foaming systems can further complicate their predictability. This presentation will focus on the "main" towers, which are typically the first separation unit to handle the effluent streams from the reactors, crackers or furnaces, in refining and petrochemical processes. Several techniques for dealing with the unpredictable problems are presented, with emphasis on monitoring of tower conditions, selection of tower internals, pre-treatment of re-cycle streams.

Effect of Recycle Streams on the Closed Loop Dynamics of Thermally Coupled Distillation Sequences

Josué Santos-Méndez and Salvador Hernández*

Universidad de Guanajuato

Facultad de Química

Noria Alta s/n

Guanajuato, Gto., 36050, México

Abstract

The effects of thermal links on the closed loop dynamics of thermally coupled distillation sequences for the separation of quaternary mixtures have been studied by using rigorous dynamic simulations. The introduction of thermal links in conventional distillation sequences can lower the energy consumption up to 40% without additional control problems in the corresponding thermally coupled distillation sequences. In some cases, the thermally coupled distillation sequences outperformed the dynamic behaviour of the conventional distillation sequences for set point tracking. This result is very important to establish that thermally coupled distillation options not only can have significant energy savings but also good dynamic properties.

Keywords: Thermally coupled distillation, energy savings, dynamic behaviour

*Corresponding author, E-mail: hernasa@quijote.ugto.mx (S. Hernández).

1. Introduction

It is well known that conventional distillation sequences consume large amounts of energy in the reboilers; as a result, researchers are interested in developing distillation schemes that can reduce both energy requirements and capital costs. One choice of great interest is the thermally coupled distillation sequences (TCDS), which can lower the energy consumption around 30% in contrast to conventional distillation trains for the separation of ternary mixtures [1,2,3,4] and also have proper dynamic properties [5,6,7]. It has been explained in the works of Triantafyllou and Smith [1] and Hernández *et al.* [8] that the TCDS options can reduce the energy consumption because they do not have re-mixing, in contrast to conventional distillation sequences. The re-mixing can be explained if one considers the separation of a ternary mixture (A,B,C) by the conventional direct distillation sequence. In the first column of the direct distillation sequence, component A is obtained as overheads and components B and C are obtained as bottoms product which is separated in the next column. If the composition profile of the intermediate component B in the liquid phase of the first column is analysed, the composition of component B increases below the feed stage until a maximum and then the composition decreases. This is known as re-mixing and is associated with higher energy consumption because in order to re-purify the mixture, additional energy will be required in the next column. Energy savings are obtained with the introduction of optimal values of the flowrates of the recycle streams. Also, the introduction of a vapour or a liquid recycle stream eliminates a reboiler or a condenser respectively.

When quaternary mixtures (A,B,C,D) are considered, there are five conventional distillation sequences (Figure 1) as reported by King [9]. At least five TCDS options can be obtained by the introduction of thermal links as shown in Figure 2. The main objective of this work is to compare the closed loop dynamic of the optimal conventional and thermally coupled distillation sequences. The dynamic behaviour is compared in terms of the integral of the absolute error (IAE).

2. Steady State Design of Conventional and Complex Distillation Schemes

The design and optimisation methods for the conventional distillation sequences of Figure 1 are well known and can be implemented by using process simulators like Aspen Plus. The retrofit method used in this work in order to obtain TCDS options (Figure 2) with minimum energy consumption is very simple and only requires the introduction of recycle streams between the distillation columns. For example, for the direct distillation sequence, the reboilers in the first and second columns are eliminated by two vapour recycle streams. All the heating needed is supplied in the reboiler of the third column as shown in Figure 2a. When the indirect distillation train is considered, the condensers in the first two columns of the sequence are replaced by two liquid recycle streams, and the total reflux required is introduced in the last distillation column. The other three distillation sequences use liquid and vapour recycle streams (Figures 2c-2e). It is important to state that in order to guarantee optimal energy consumptions, the interconnecting streams must be varied until the minimum energy requirements are obtained. The energy-efficient designs of the conventional and thermally coupled schemes were obtained in Aspen Plus 11.1^{TM}.

3. Dynamic Study and Case of Study

The optimum conventional and thermally coupled distillation sequences were subjected to a study of their dynamic responses under the action of feedback PI controllers. Changes in the set points were considered and the optimal responses were chosen in terms of their IAE values. This study was completed in Aspen Dynamics 11.1^{TM}. The separation of a mixture of n-pentane, n-hexane, n-heptane, n-octane with flows of 13.62, 9.08, 9.08 and 13.62 kmol/h respectively was studied. Compositions of 0.98, 0.97, 0.97 and 0.98 in mole fraction for products A, B, C and D respectively were assumed. The pressures in the columns were set in order to use cooling water in the condensers. The pairings in the control loops were selected according to practical aspects, the overheads compositions were tied to the reflux rates (Figure 3 and 4) and the bottoms compositions were controlled by manipulating the heat demands in the reboilers.

4. Results

In the first part of the study, energy-efficient designs were obtained for the conventional and thermally coupled distillation sequences. The design and optimisation of the conventional distillation sequences followed standard methods in the process simulator Aspen Plus 11.1^{TM}, but for the case of the thermally coupled distillation schemes, the optimisation task included a complete search in the recycle streams. In the optimised design of the conventional distillation sequence, thermal links were introduced as described above and the minimum energy consumption was detected in the thermally coupled distillation sequence. For the case of study considered in this work, Figure 5 shows the search for the optimal value of heat duty (QR); we can observe that the energy consumption depends strongly on the values of the vapour recycle streams (VF1 and VF2 in Figure 5). In this case, savings in energy of up to 40% were obtained in the thermally coupled distillation sequences in contrast to conventional distillation sequences (Table 1). Also, all the thermally coupled distillation sequences demanded less energy consumption in comparison to the corresponding conventional distillation sequence.

Regarding the dynamic responses under the action of feedback controllers for changes in the set points, Table 2 shows the IAE values for a positive set point change in each product composition. The TCDS I outperformed the dynamic responses of the CDS I for set point changes in components A and B as shown in Table 2, but for components C and D the CDS I presented better values of the IAE. When a negative set point change of the same magnitude was implemented in each composition, a similar result was obtained. The CDS I presented the best dynamic behaviour according to its lower values of the IAE (Table 2) in comparison to the TCDS I for components B and D; on the other hand, when the set points of compositions of the products A and C were perturbed, the TCDS I presented better dynamic responses than those obtained in the CDS I.

Figures 6 and 7 present the dynamic responses for the CDS I and TCDS I respectively when a positive change of 0.005 was implemented in the product composition A. It can be observed that the dynamic response of the TCDS I is better than that obtained in the CDS I. The dynamic response reached the new steady state faster. When a positive set point change in component D was considered, Figures 8 and 9 show the dynamic behaviour and for this case the CDS I presented better dynamic response than the TCDS I, but it is important to say that both types of distillation sequences can achieve the change in the set point in a short period of time. It can be stated that both types of distillation sequences present similar dynamic behaviour but the integrated distillation sequences have significant energy savings in contrast to the original uncoupled distillation sequences. This of course is an incentive for the retrofitting of conventional distillation sequences through the use of thermal links. Also, in the design of new plants, thermally linked distillation sequences can have important savings in both energy and capital costs, and the savings are obtained without creating operational and control problems.

5. Conclusions

The effect of recycle streams on energy consumption and dynamic responses was studied for the separation of a quaternary mixture of hydrocarbons. The five conventional distillation sequences were subjected to retrofit by introducing mass recycle streams between the columns. Energy savings of up to 40% were achieved when the recycle streams were optimised. The dynamic responses of the thermally coupled distillation sequences were as good as those obtained in the conventional distillation sequences, and in some cases the dynamic responses of the thermally coupled distillation sequences were better than those of the conventional distillation schemes. These results establish that the energy savings can be achieved without introducing additional control problems as it might be expected originally.

Acknowledgement

J. Santos-Méndez was supported through a scholarship from the Universidad de Guanajuato, México, during the development of this project.

References

1. Triantafyllou, C., and Smith, R. (1992). The Design and Optimisation of Fully Thermally Coupled Distillation Columns. *Trans. Inst. Chem. Eng.*, **70**, 118-132.

2. Hernández, S., and Jiménez, A. (1996). Design of Optimal Thermally-Coupled Distillation Systems Using a Dynamic Model. *Trans Inst. Chem. Eng.*, **74**, 357-362.

3. Hernández, S., and Jiménez, A. (1999a). Design of Energy-Efficient Petlyuk Systems. *Comput. Chem. Eng.*, **23**, 1005-1010.

4. Caballero, J. A. and Grossmann, I.E. (2001). Generalized Disjunctive Programming Model for the Optimal Synthesis of Thermally Linked Distillation Columns. *Ind. Eng. Chem. Res.*, **40**, 2260-2274.

5. Abdul Mutalib, M.I., Zeglam, A.O., and Smith, R. (1998). Operation and Control of Dividing Wall Distillation Columns. Part II: Simulation and Pilot Plant Studies Using Temperature Control. *Trans. Inst. Chem. Eng.,* **76**, 319-334.

6. Hernández, S., and Jiménez, A. (1999b). Controllability Analysis of Thermally Coupled Distillation Systems. *Ind. Eng. Chem. Res.*, **38**, 3957-3963.

7. Jiménez, A., Hernández, S., Montoy, F.A. and Zavala-García, M. (2001). Analysis of Control Properties of Conventional and Nonconventional Distillation Sequences. *Ind. Eng. Chem. Res.*, **40**, 3757-3761.

8. Hernández, S., Pereira-Pech, S., Jiménez, A., and Rico-Ramírez, V. (2003). Energy Efficiency of an Indirect Thermally Coupled Distillation Sequence. The Can. J. Chem. Eng., in press.

9. King, C. J. (1980). *Separation Processes*, 2nd Edition, McGraw-Hill, NewYork, USA.

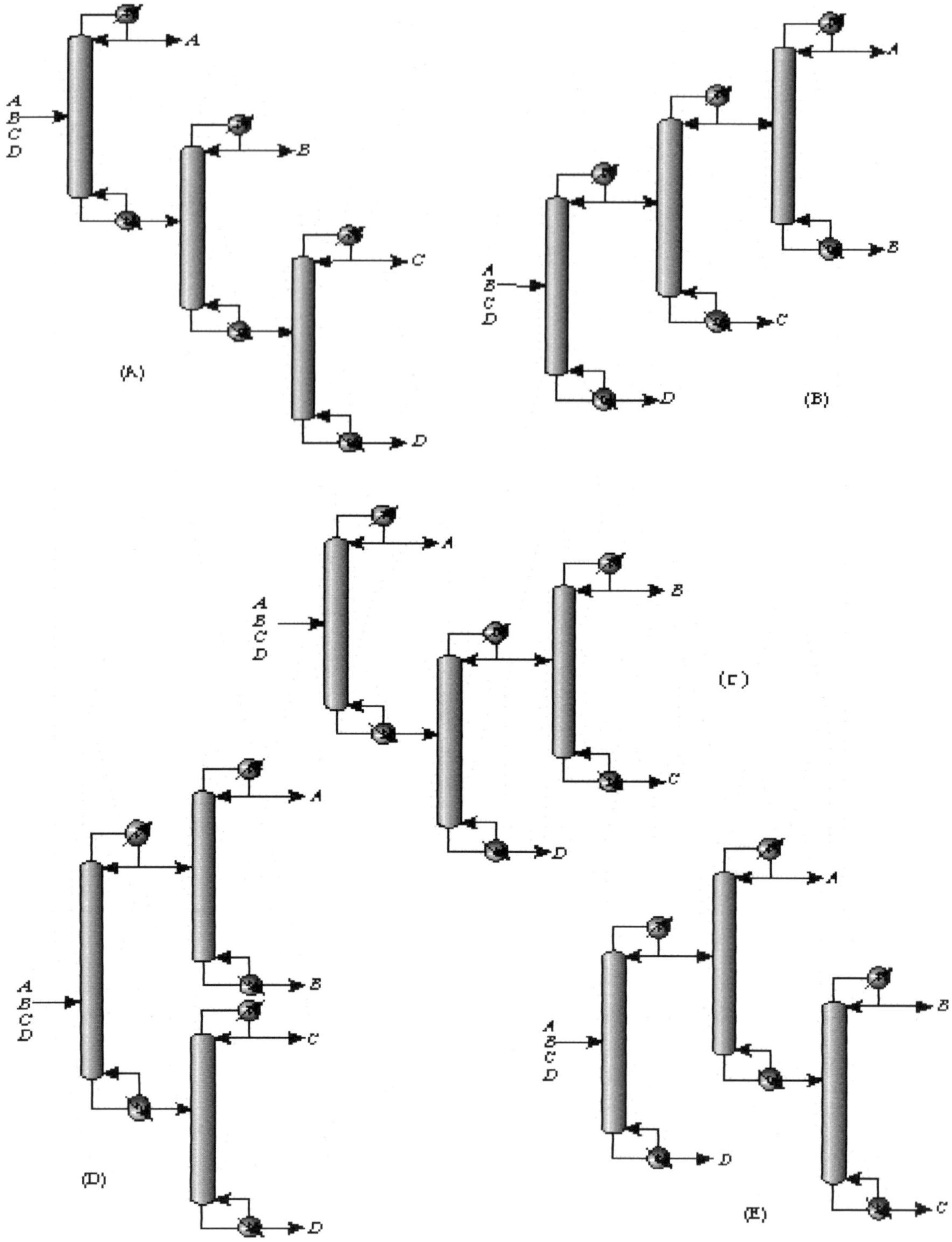

Figure 1. Conventional distillation trains. (a) CDS I direct, (b) CDS II Indirect, (c) CDS III, (d) CDS IV, (e) CDS V.

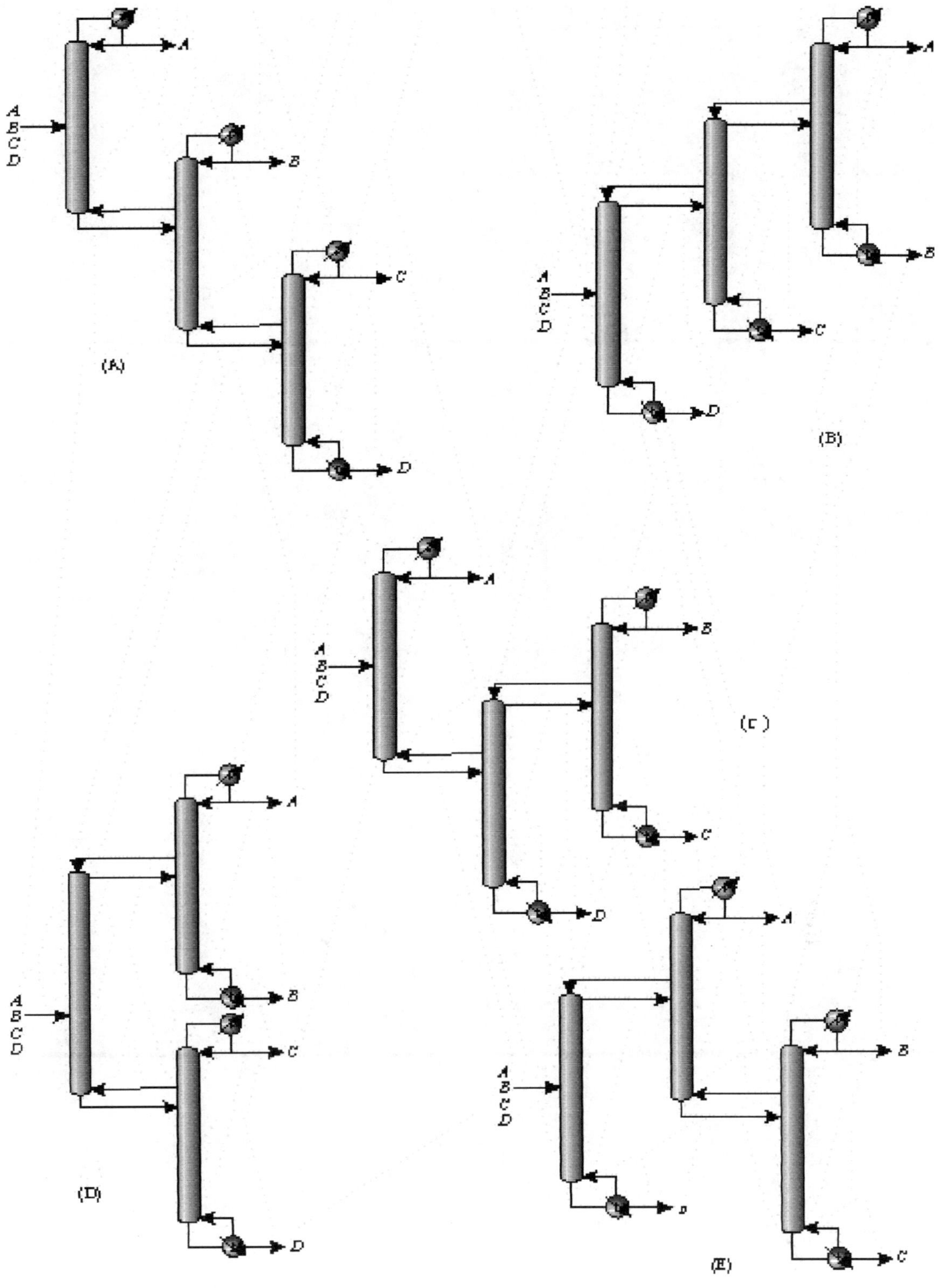

Figure 2. Thermally Coupled Distillation Sequences. (a) TCDS I direct, (b) TCDS II Indirect, (c)TCDS III, (d) TCDS IV and (e) TCDS V.

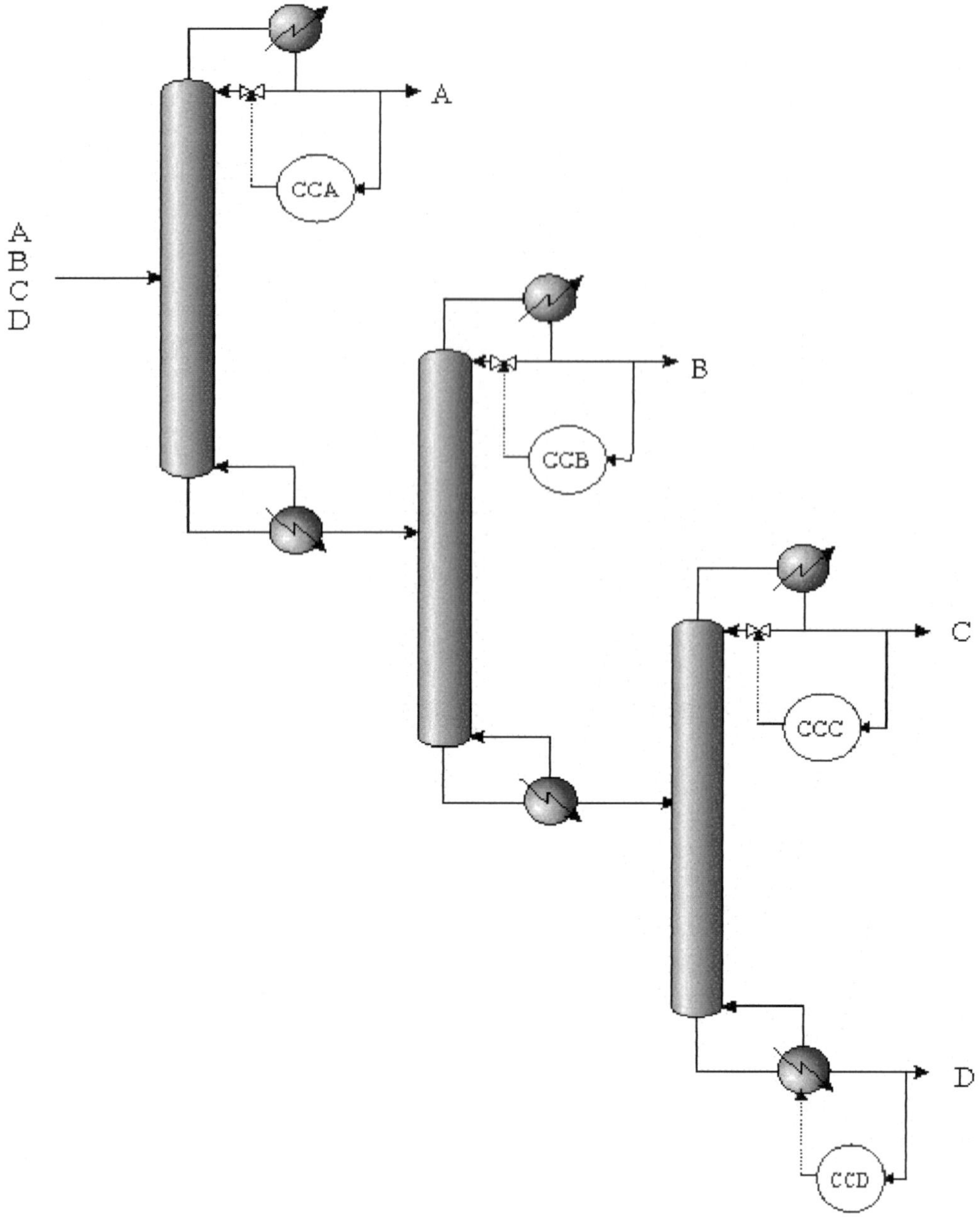

Figure 3 Control loops in the CDS I.

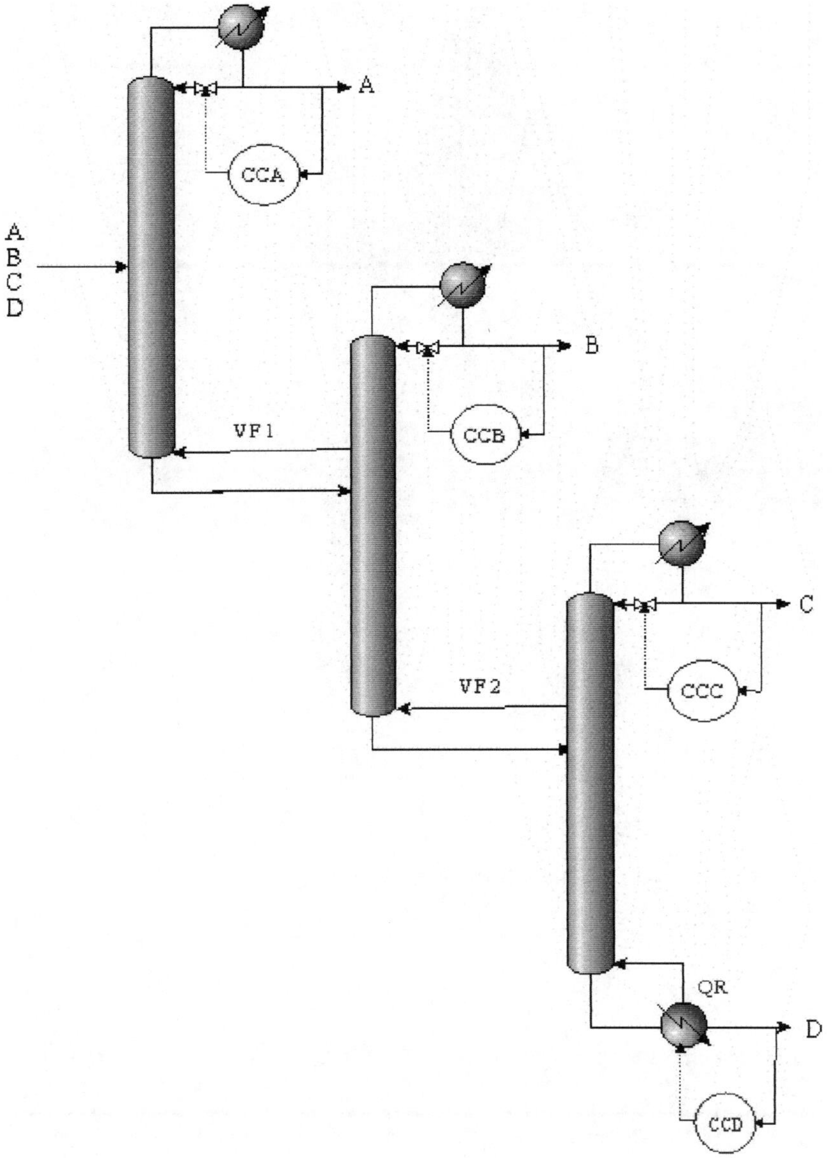

Figure 4 Control loops in the TCDS I.

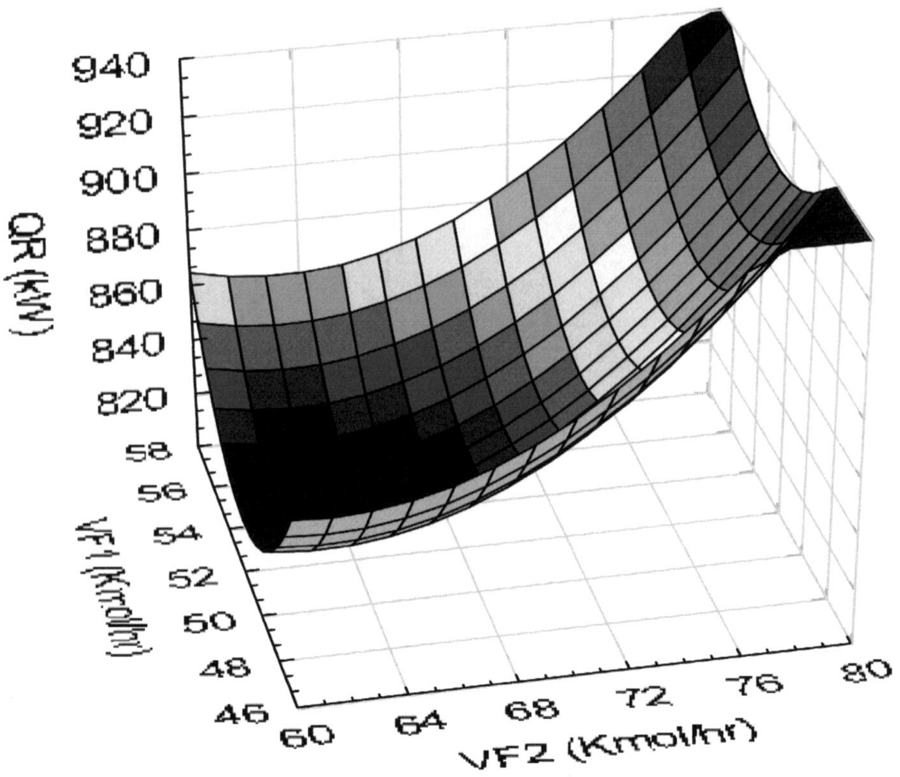

Figure 5 Search for the optimum energy consumption in the TCDS I.

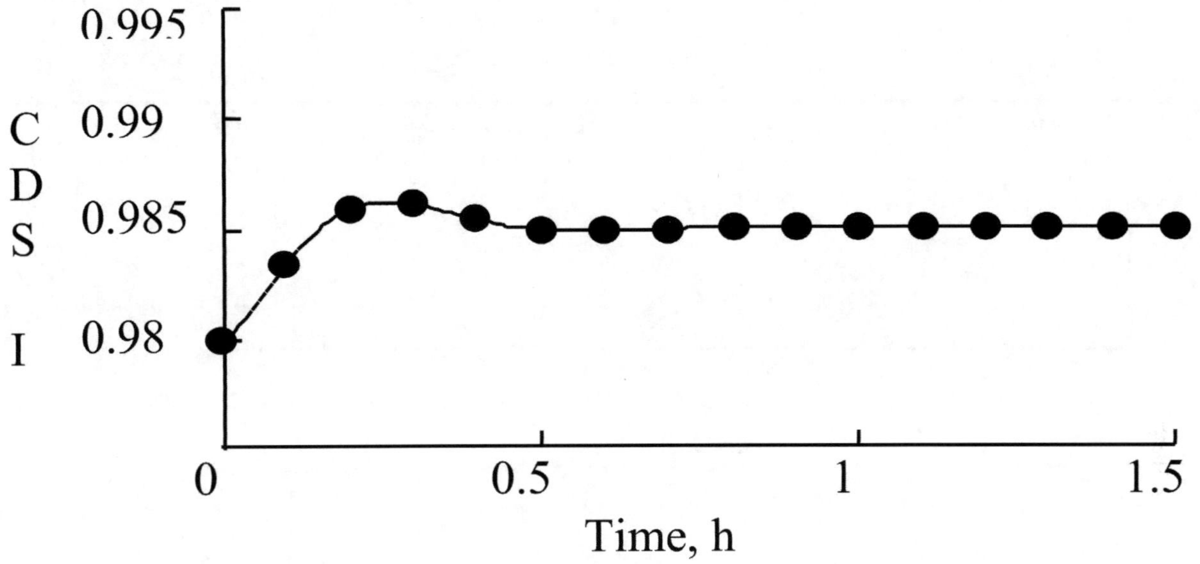

Figure 6 Dynamic response in the CDS I for a positive set point change in component A.

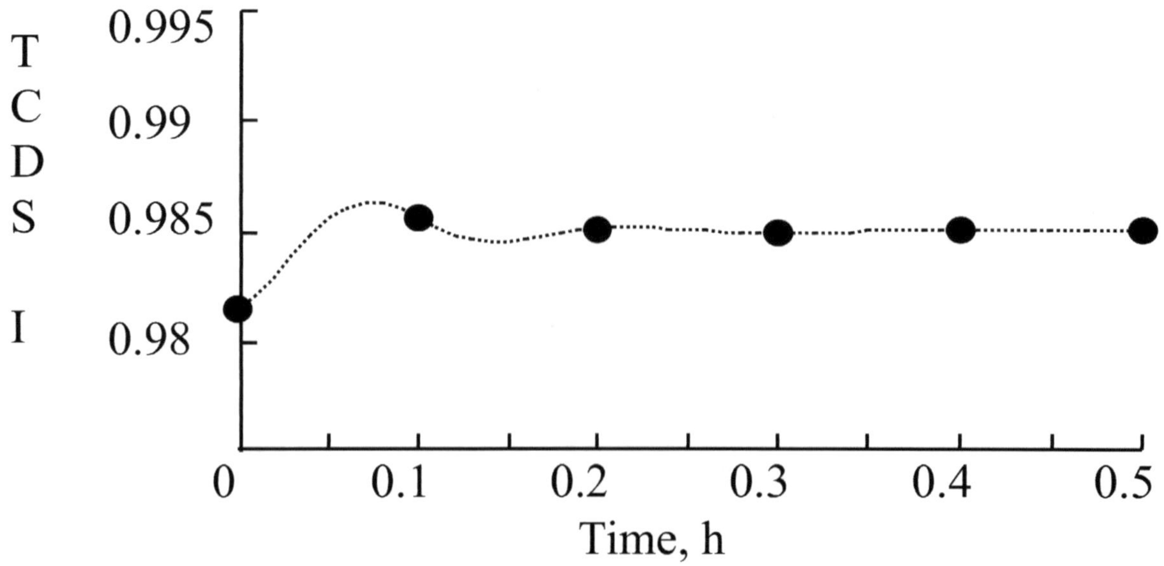

Figure 7 Dynamic response in the TCDS I for a positive set point change in component A.

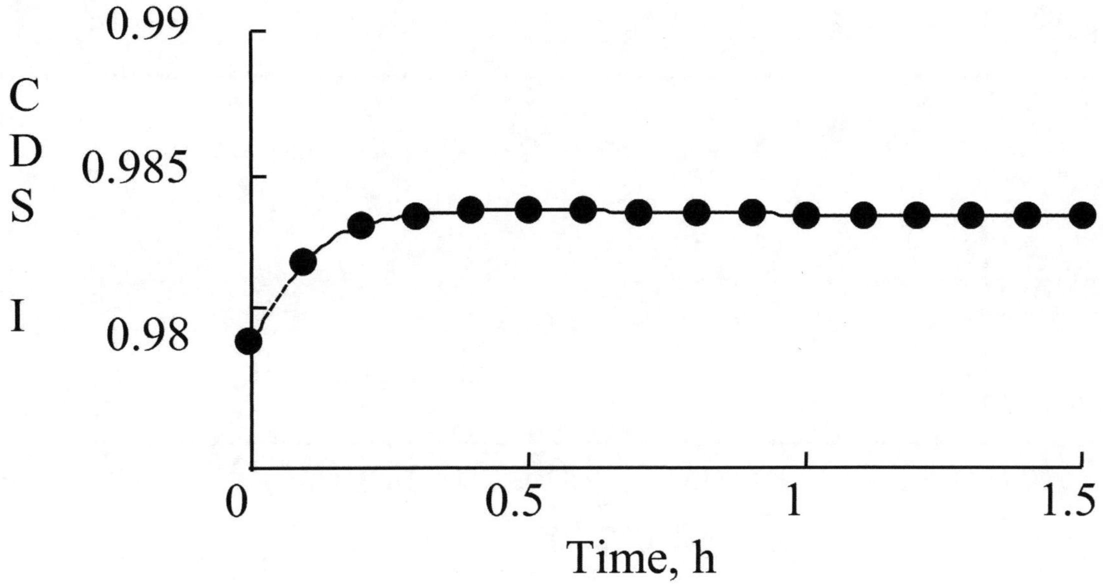

Figure 8 Dynamic response in the CDS I for a positive set point change in component D.

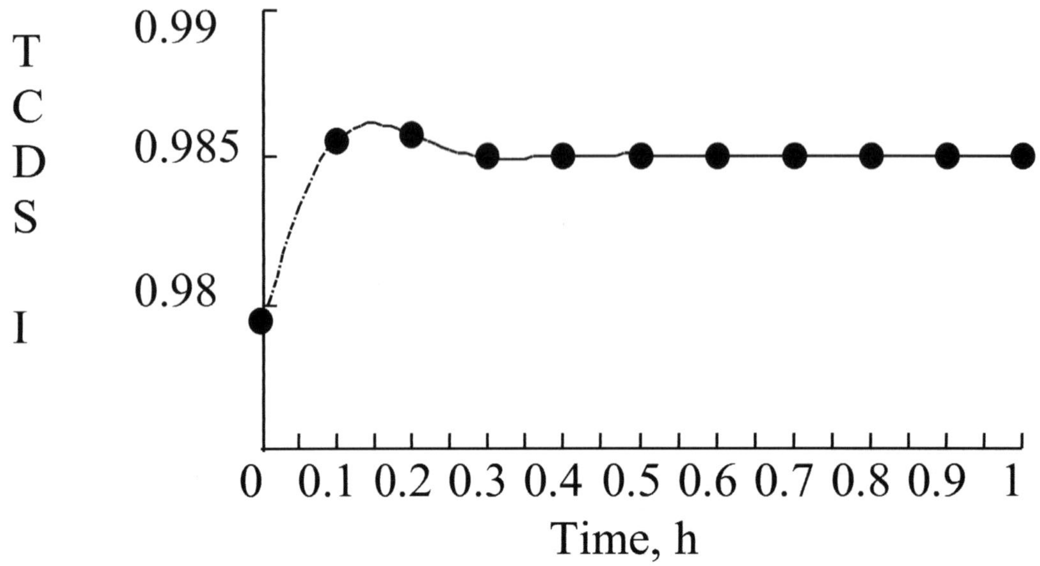

Figure 9 Dynamic response in the TCDS I for a positive set point change in component D.

Table 1 Optimal energy consumptions (kW) for the conventional and complex distillation sequences.

CDS I	CDS II	CDS III	CDS IV	CDS V
1122.5	1188.5	1503.2	1174.9	1780.9
TCDS I	TCDS II	TCDS III	TCDS IV	TCDS V
813.8	921.2	1106.5	836.5	942.4

Table 2 IAE values for positive and negative changes in each set point composition.

Distillation sequence	Product stream			
	Positive change in set points			
	A	B	C	D
CDS I	0.0143109	0.0142759	**0.0001042**	**0.0004593**
TCDS I	**0.00494638**	**0.0005048**	0.0049209	0.0099678
	Negative change in set points			
CDS I	0.00037819	**0.00023645**	0.0007281	**0.0000745**
TCDS I	**0.00026217**	0.00107824	**0.0000397**	0.0003016

AUTHOR INDEX

AUTHOR INDEX

American Institute of Chemical Engineers
3 Park Avenue
New York, NY 10016-5991

ISBN: 978-1-60423-355-1